S0-BZR-656

Millicoma

Location of the Millicoma Tree Farm

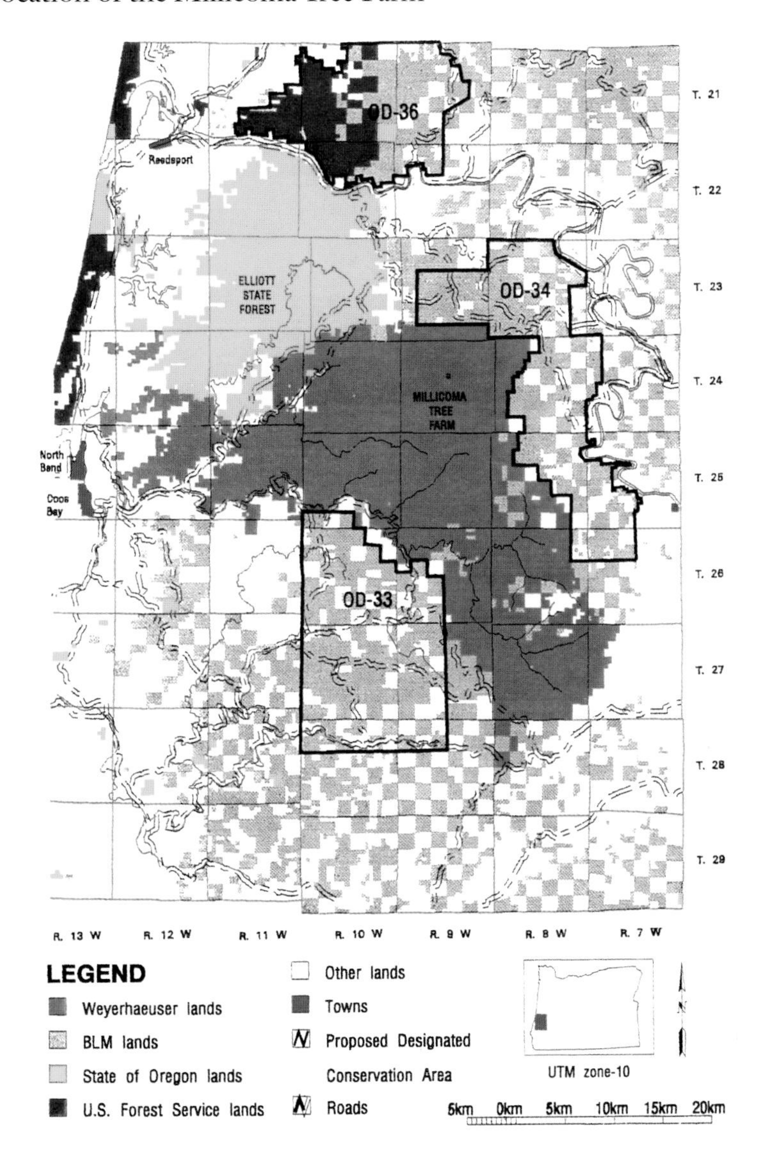

Millicoma Tree Farm location map. Source: Beak Consultants Inc., Environmental Assessment for the Proposed Issuance of a Permit for Incidental Take of the Northern Spotted Owl: Millicoma Tree Farm, Weyerhaeuser Company, Coos and Douglas Counties, Oregon. Prepared for the U.S. Department of the Interior, Fish and Wildlife Service. November 16, 1994.

Forest Cover Types on the Millicoma Tree Farm

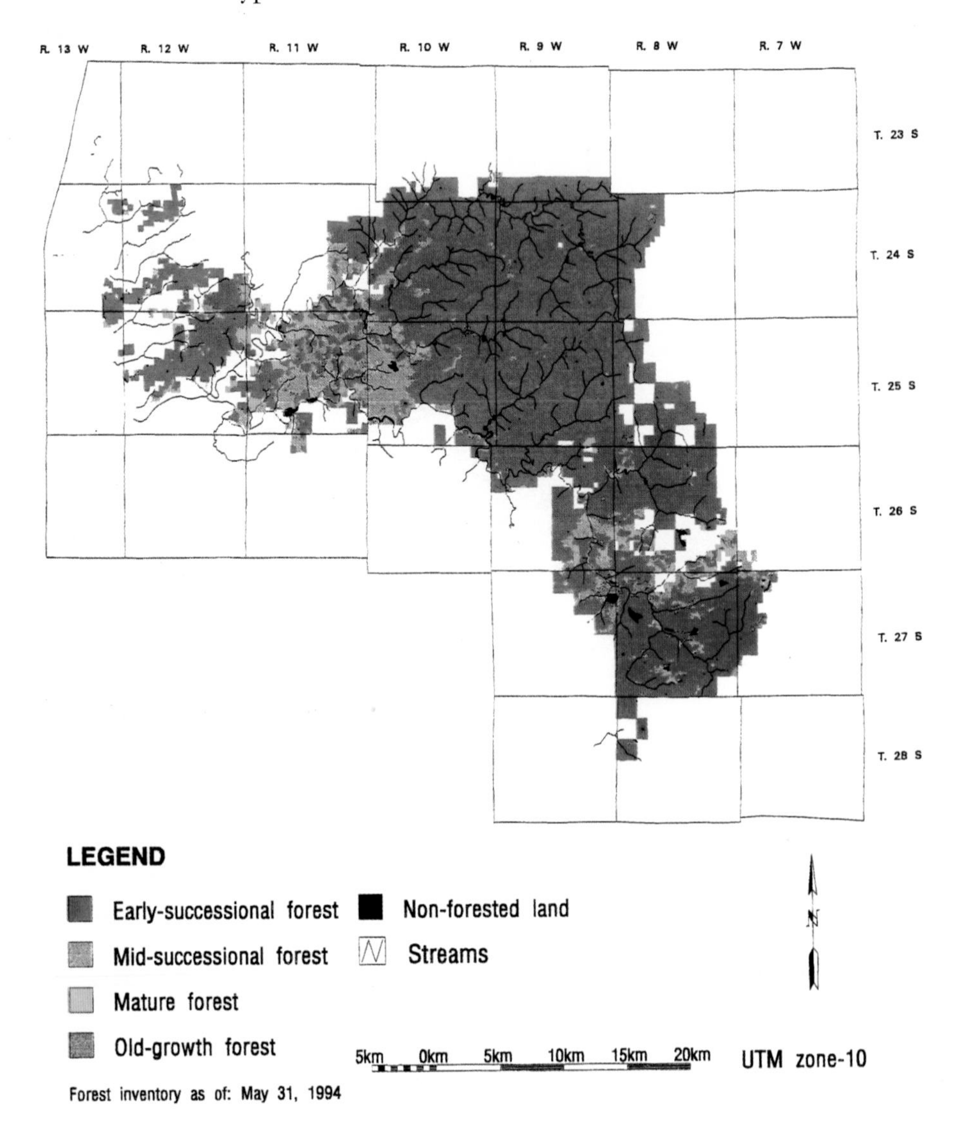

Forest cover types on the Millicoma Tree Farm. Source: Beak Consultants Inc., Environmental Assessment for the Proposed Issuance of a Permit for Incidental Take of the Northern Spotted Owl: Millicoma Tree Farm, Weyerhaeuser Company, Coos and Douglas Counties, Oregon. Prepared for the U.S. Department of the Interior, Fish and Wildlife Service. November 16, 1994.

Arthur V. Smyth

Millicoma

BIOGRAPHY OF A PACIFIC NORTHWESTERN FOREST

FOREST HISTORY SOCIETY Durham, North Carolina 2000

© 2000 Forest History Society
Printed in the United States of America
All rights reserved

The Forest History Society is a nonprofit, educational institution dedicated to the advancement of historical understanding of man's interaction with the forest environment. It was established in 1946. Interpretations and conclusion in FHS publications are those of the authors; the institution takes responsibility for the selection of topics, the competence of the authors, and their freedom of inquiry.

Contact the Forest History Society at
701 William Vickers Avenue,
Durham, NC 27701-3162
(919) 682-9319; fax (919) 682-2349
http://www.lib.duke.edu/forest/

This publication was made possible by contributions from George H. Weyerhaeuser, Sr. and the Weyerhaeuser Company Foundation.

Designed by Amy Ruth Buchanan

On the cover: The 180-year-old Millicoma Forest, Oregon (Weyerhaeuser Archives photo).

Fold-out map in back of book: Hunter's map of the Millicoma Forest, provided by the Weyerhaeuser Company, 1996.

Library of Congress Cataloging-in-Publication Data appear on the last printed page of this book.

To my wife, Irene

Contents

Foreword by Charles E. Twining IX

Preface by Daniel B. Botkin XI

Acknowledgments XVII

Introduction 1

1 / Whose Woods These Are 6

2 / What Do We Have Here? 16

3 / Harvesting the Forest—Early Planning 28

4 / The Early Development 36

5 / The Forest Gets a Name 43

6 / The Logging Begins 48

7 / The People 54

8 / Chasing the Beetle 59

9 / Changes in the Forest, Changes in the People 65

10 / Wind, Water, and Fire 75

11 / High Yield Forestry 82

12 / A Changing Society 95

13 / More Than Trees 100

14 / The Transition 106

15 / The Last Raft 109

16 / The Owl and the Millicoma 113

17 / A New Look 120

18 / Habitat Conservation Area 124

19 / The Millicoma Forest: Today and Tomorrow 129

Epilogue 139

Glossary 141

Index 143

Foreword / **by Charles E. Twining**

This insider's story will inform and entertain relative newcomers to the subject. While it is the rare individual who fails to find beauty and perhaps spiritual solace in the presence of trees, most of us view forests from a visitor's perspective. As significant as we may find such visits, the experience is very different for those who have spent their careers in the woods and the forest products industry.

Few of us have started out working with the nuts and bolts of an operation and ended dealing with national policies that affect how those nuts and bolts are used and fit together. Art Smyth has, and the pages that follow describe his advancement over the years. He began his education in the western woods following graduation in 1941 from the University of Michigan School of Forestry and Conservation. His first assignment with Weyerhaeuser was a staff position at the Clemons Tree Farm, subsequently certified as the nation's first. In short, Art commenced his career coincident with new thinking regarding forests and forestry.

He was next assigned responsibilities in the Millicoma Forest at Coos Bay, Oregon, and this account will speak largely to that experience. Art has thought of this story as the biography of a forest, and so it is. This is no ordinary biography, however, not only in terms of the subject, but also insofar as the story is never-ending. Certainly Millicoma Forest had its beginnings, but thus far the forest continues, as healthy and productive as ever.

Some readers may be a bit surprised that a forester, at least one who is involved in growing trees for ultimate harvest, could view them with such genuine feeling and respect. Those who have come to know foresters will not be surprised. Fred Conant, one of Weyerhaeuser's early cruisers, admitted as much in a 1925 letter: "Many times I have stopped in the woods, alone, and taken my hat off to a fine tree—the only thing that gives me a genuine impulse to do that." And years later, in a 1979 interview, Ed Heacox, Weyerhaeuser's chief forester, agreed. "I've been in the woods business and forestry business all my life and yet when I'm in the woods and I see a big tree go down, there's a little twinge always." Art Smyth would understand.

Although most know Art as the company's lobbyist in Washington, D.C., Art's effectiveness in that position stemmed from his long experience in the forestry and production end. Prior to his move to "the other Washington," Art had served as raw materials manager at Longview, Washington, then Weyerhaeuser's largest manufacturing facility. Thus, in his own words, he went "from counting logs on the Columbia to rolling logs on the Potomac."

Most important, he knew of what he spoke; and he spoke directly and honestly, always in good humor, a trait his colleagues appreciated. Art would never get carried away by the trappings and opportunities of working with government officials. Perhaps it was because of his ability to recall earlier times. More likely, it simply reflected Art's friendly and unassuming personality.

So, I invite the reader to pull up a chair, or if possible a log, and join Art on his mostly merry adventure.

Charles Twining
Federal Way, Washington

Preface / **by Daniel B. Botkin**

In recent years, forests have captured the popular imagination, and wilderness in forests has become a positive social value. Major public debates about forest issues include: What is the importance of old growth forests? Is old growth the *only* important and valuable state of a forest? Is clearcutting ever justified, and if so, by what methods, how often, and how large? What is the natural role of fire in forests? Is prescribed burning—fires intentionally lit by people—ever justified? Is it better or worse to reduce excess fuel—by burning or by logging dead wood—to bring forests back to a pre-settlement level of fuel loading? What is required to conserve endangered species that live or depend on forests? How might we maintain employment in forest-related industries, provide forest products worldwide, yet sustain forests? For these questions, there are many debates but few clear answers.

Forests of the Pacific Northwest are central to this public concern. The immense trees of this region, reaching more than two hundred feet high and extending more than six feet in diameter, capture our imaginations as a realization of nature's grandeur. Many issues in this region concern wildlife that inhabit the forest, such as salmon, the spotted owl, and the marbled murrelet.

Accompanying the increase in popularity of forests is a set of common beliefs that have entered our folklore, whose origin traces back to Greek and Roman myths about the balance of nature. Among these beliefs are that undisturbed forests, including those that existed in the Pacific Northwest of North America prior to European colonization, were all old growth, and that these were magnificent and beautiful because they were free of human influence.

During most of the twentieth century, ecological science reinforced this folklore, claiming that every forest grew from a clearing to a constant state, called the climax forest, or what we refer today as "old growth." This climax forest was supposed to contain the greatest amount of organic matter, have the greatest biological diversity, and the greatest store of biologically essential chemical elements compared to any other stage in forest development. When disturbed, old growth forests were supposed to follow a regular progression, called forest succession, to return to exactly the same climax state.

These beliefs have been expanded and enlarged in our folklore. Forests, in a way, have once again become sacred groves, just as they were to the ancients, and believed to be best without human influence, best if only visited and untrammeled by human beings. This folklore feeds back, through public pressure and through the acceptance of these beliefs by activist groups, to create a body of laws and regulations that determine how we can affect forests.

But in the last quarter of the twentieth century, the basis for these common beliefs became discounted by growing scientific evidence of the dynamics of nature. Nature, it became clear, always changed. Constancy was not the normal state of a forest, and the greatest biological diversity in a forest did not necessarily occur in the oldest stands. So the idea of a steady state of nature—in general and for forests in particular—was shown to be incorrect according to the best available scientific information. Forests were instead recognized as dynamic, as changing over many time scales.

Ecologists also learned during the latter part of the twentieth century that many species were adapted to forest change. Some, like the endangered Kirtland's warbler that nests only in jack pine woodlands, which in turn depend on fire, are adapted to and require forest disturbances. Some species, like jack pine and other pines, produce an abundance of fuel. The lower branches of jack pine tend to remain on the tree and do not readily decay. In this way, such trees promote fire, and this evolutionary adaption makes it possible for them to "win" in competition with trees that can better tolerate deep shade but are not tolerant of fire. Not only is change natural in forests, it is necessary for such species.

Another finding of late-twentieth-century sciences is that often forests that we enjoy and had believed to be "virgin"—untouched by human actions—turned out to be beautiful and pleasing to visit as a direct result of human activities in combination with environmental change and internal forest dynamics. This is true for many forests that were subjected to periodic burning by the Indians. These light, frequent fires created a forest of large, mature trees, widely spaced, and with relatively few shrubs; these magnificent forests are the kind believed to be virgin.

But ideas are powerful and diehard. In spite of the change in scientific understanding of forests, the folklore of the perfection of steady-state forests persists. Bad forest practices, resulting sometimes in huge clearcuts, reinforce the old myths. Examples of poor forest harvest practices are interpreted to be proof that *all* forest harvests are and must be destructive, that people are incompatible with forests.

As long as there has been civilization, people have had a love-hate relationship with forests. Forests provided the major fuel and construction material for early civilization. (It was only at the turn of the twentieth century that coal surpassed wood as the major fuel source in the United States.) People feared the darkness of forests, welcomed cleared and opened land, which felt safer, but at the same time civilization depended on forest products.

Although harvesting of timber has a long history, it continues to be an ongoing, learning process. We are just beginning to understand what may be required for truly sustainable forest practices. The scientific understanding of forests has proven a difficult endeavor, and we still know little about what makes for sustainable forests, whether they are subject to human harvest or not.

How do we change our ideas, so that our beliefs about forests become consistent with our knowledge, and so that our laws can be based upon the best scientific information rather than folklore? In his life and works, Henry David Thoreau tackled these issues about nature in an interesting and specific way. First, he sought to learn directly from his own experience. Second, if he could not obtain that experience himself, he sought to listen to, or read the works of, someone who had such experiences. Third, he sought the details of a situation, and, examining these, let the generalizations arise by themselves. Fourth, he used the experiences of others as hypotheses—as starting points for his own investigations, stimulated by the most interesting ideas of others.

When it came to nature, Thoreau relied on those who had experience with, and depended for their livelihood on, natural resources. "Fishermen, hunters, woodchoppers, and others, spending their lives in the fields and woods, in a peculiar sense a part of Nature themselves," he wrote in *Walden,* "are often in a more favorable mood for observing her than philosophers or poets, who approach her with expectation."[1] By "expectation" I believe he meant with common beliefs, myths, and ideologies.

With the present concern with forests and forest practices, Arthur Smyth's book—written by a practicing forester and based on many forms of knowledge, including his own direct experiences, oral histories obtained from others experienced in the woods, and from scientific studies—follows the method of Thoreau and allows us to do so as well. The way out of the box of oversimplified generations is to learn stories about specific forests, staying faithful to the facts of history and science. If we cannot experience the history of a specific forest for ourselves, then a book like Arthur Smyth's provides the details that allow us to learn that history.

In his book, Arthur Smyth tells us about the Millicoma Forest, which, he writes, was "several hundred thousand acres of some of the richest privately owned forestland in North America." He tells us an unfolding drama—human as well as ecological and geological. Rather than a simple situation with people in white hats and black hats, with a clear single truth, and clear good and evil, we discover through Smyth's personal experiences, through oral history, through written history, and through the history that trees and forests tell themselves, an intricate story full of characters and controversies, hard work and humorous accidents, warts and worries, but most important, about how the forests and societal attitudes have changed over time.

Smyth documents the history of the Millicoma Forest before the arrival of Europeans. A million years ago, the land was under the ocean. Studies of cores from many trees, in which he participated, show that the forest must have been subjected to a large fire in 1765 and affected by storms since then.

Against this geological history, the cutting of forests in the Pacific Northwest is relatively recent—mostly completed after World War II. And the first cutting of the forest occupied a surprisingly brief time—about forty years, with the big mill closing down in 1989. During this forty-year period, great changes occurred in the American perception of forests. At first, timber was viewed as a commodity, if people paid attention to forests at all.

Arthur Smyth has had extensive personal experience in this forest, beginning with the task of determining the species composition, rate of growth of

1 From Thoreau's *Walden* p. 198. Quoted in Botkin, *No Man's Garden: Thoreau and a New Vision for Civilization and Nature* (Washington, D.C.: Island Press, 2000).

trees during the previous fifty years, and the mortality rate of trees. Detailed measurements—unusual in forestry—were done with two man crews running lines a mile apart, measuring a quarter acre every six hundred feet. They obtained cores from almost 1,500 trees to determine growth rates and fire history. This was rugged outdoor work—camping twenty-eight days at a time in the rainy Pacific Northwest.

The stories he tells suggest that the harvest of the Millicoma not only occurred rapidly but also, in retrospect, without sufficient concern for the effects of large-scale clearing on the public's perceptions or the community's concerns. During the short time of the first cut of this forest, societal attitudes changed greatly about wildlife and fish in these forest lands. When logging started just after World War II, bounties of cougars and other predators had been in existence in Oregon for a century. With the coming of the environmental era—beginning in the 1960s—and the realization in that decade that as few as two hundred cougars might still exist in the entire state, social attitudes changed; predators became protected, and the cougars have since recovered considerably.

The Weyerhaeuser Company, owner and operator of the Millicoma Forest, began this era priding itself—and advertising itself—as the "tree-growing company," an idea that seemed to function in the early days of the logging of the Millicoma. But by the 1970s, this idea was out of step with social attitudes. The biggest change in social attitudes about the Millicoma Forest, and other forests of the Pacific Northwest, began in that decade with concern over the spotted owl. Since then, our society has undergone a series of major changes in the appreciation of forests and forest-dwelling animals. While in 1950 relatively few Americans may have been concerned strongly about forests, by the 1980s forests had become an important environmental value, and most people appreciated forests for their landscape beauty, biodiversity, and clean streams, full of salmon, steelhead and other fish. People did not like to see large clearcuts. As Smyth points out, tree farms were seen as sterile monocultures. These new societal attitudes have led to changes in forest practices. In 1991, the U.S. Fish and Wildlife Service listed critical habitat for the spotted owl to include 11.6 million acres, including 3 million acres of private lands.

And so perhaps Henry David Thoreau's approach was the right one. Perhaps the answers to the questions I have posed will be best answered by careful consideration of detailed history of forests and of their use. This is the great value of Smyth's fascinating saga of one of Oregon's most productive and intensively harvested forests. In the final analysis, what Smyth's book seems to suggest is that, to deal with and solve our environmental problems about forests, we should understand and learn from the details of forest history—obtained from many different avenues—and that we need to see the use of forests against the panorama of changes in human values and changing scientific knowledge.

Daniel B. Botkin
Santa Barbara, California

Acknowledgments

I have wanted to tell this story for some time. The controversies that have swirled around the forests of America have been fueled by the deep-seated values many people have about nature. In most cases, these people live far from the lands that concern them. This is the story of a real forest, real people, and the forces, both human and non-human, that shaped the Millicoma Forest.

Before retiring, I spent virtually my entire career with the Weyerhaeuser Company as a forester and government relations executive. No one from the company exercised any editorial control over this manuscript; however, they were extraordinarily open with the company archives and files. Without the cooperation of Charley Bingham, who gave me access to the company archives; Mack Hogans who encouraged me; and Donnie Crespo, the company archivist who helped me immeasureably, I could not have written this book. Bruce Beckett was most helpful in supplying information on the Habitat Conservation Area process. Donna Brown was an early reviewer and supplied me with the research she had done while at Coos Bay in the 1980s.

Clyde Kalahan shared valuable information, including the climatic studies of the western United States. Harry Morgan provided his taped interview. A host of retired foresters who had worked on the Millicoma provided me with their recollections, photos, and memorabilia. These include Dean Higinbotham, Herm Sommer, Jim Lavan, Ralph Sweet, Don Wickendoll, Skip Dunlap, and Hank Repetto. Al Pettey was also particularly helpful. When he retired he had become the most senior forester on the Millicoma, having spent almost his entire career there.

The present staff at the Coos Bay operations, including Jack Taylor, Timm Slater, Mary VanPulliam, and Karen Whaley, was most helpful. My special thanks to Jim Clarke, who one rainy December day guided me through the "new" Millicoma Forest on my last visit.

My thanks also to Rudolph Rosen, director of the Oregon Department of Fish and Wildlife, and his staff for supplying me the research reports on the Millicoma elk. Bob Corthell, retired department biologist, was most helpful with his information on the Matson Creek trout, as was Ken Wright, retired Forest Service entomologist, with his information on the beetle research.

Two of my sources are unfortunately no longer with us. George Staebler, the retired director of forest research for Weyerhaeuser, recently died of a heart attack. He supplied me valuable information on fire history. Royce Cornelius, although he was dying of brain cancer, took the time to answer many of my questions about his early days on the Millicoma.

The book would not have been published without the support of George Weyerhaeuser. Rick Weyerhaeuser was an early supporter and supplied the correspondence between Frederick Weyerhaeuser and his son Davis. My thanks to my publisher, the Forest History Society, and its president, Steve Anderson. My special thinks to the managing editor, Karie Kirkpatrick, who made the manuscript into the product you see.

I was a personal witness to much of this story, but obviously I was not on the deck of the Spanish frigates or around Coos Bay in the 1850s. For this historical background I depended on the original research of many scholars and historians. My references are listed at the end of each chapter. I did not use footnotes or detailed notes to document my sources. I am a storyteller, not a professional historian. For a small community, the Coos Bay area has produced many excellent local histories. The books of Orvil Dodge, Emil Peterson, Charlotte Mahaffey, and Dow Beckham were all used. I found Lionel Youst's book on the Glenn Creek homesteaders especially helpful. Charles Twining is the historian most expert on the Weyerhaeuser family, and I have used his books on George Long and Phil Weyerhaeuser extensively. My sources on the Endangered Species Act and the spotted owl controversy were principally Charles Mann and Mark Plummer's book, *Noah's Choice,* and Steven Yaffee's book, *The Wisdom of the Spotted Owl.* Much of my research on the early explorations was done in the magnificent reading room of the Library of Congress. I was thrilled to be reading the original journals of David Douglas in the elegant Rare Book Room of the library. My thanks to the many who had the vision to create such a national resource.

If I have forgotten someone who helped me, I apologize. The book is dedicated to my wife, Irene, who has encouraged me every step of the way since the day forty-four years ago we moved into House #2 at the Allegany logging terminal.

For the benefit of readers who may not be familiar with some of the words used in the story, I have included a glossary at the back of the book.

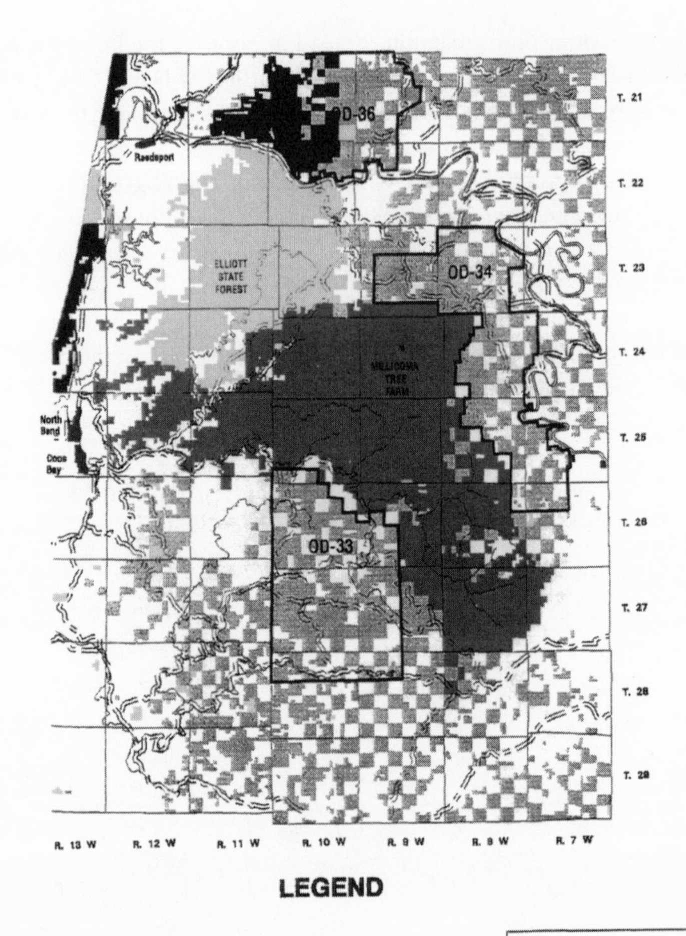

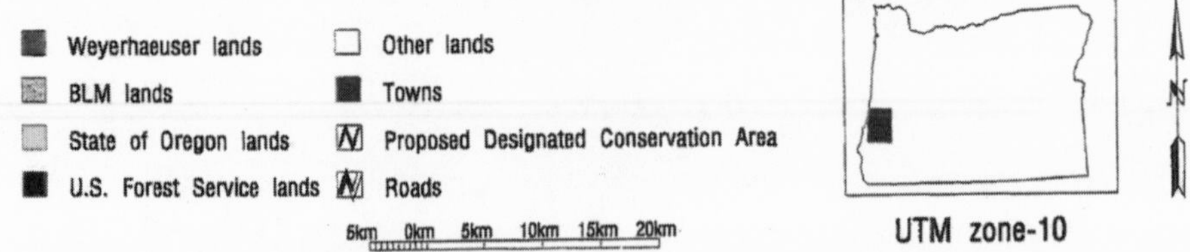

Millicoma Tree Farm location map. Source: Beak Consultants Environmental Assesments.

Introduction

This is a story of the life of a forest—how it began and how the forces of nature and the human hand changed it. No attempt has been made to identify villains or heroes. Rather, this is what happened to a landscape over several generations. Readers can put their own values on the events. This is the story of a human relationship with a landscape. A *biography* of a forest? Yes, a forest breathes, grows, and reproduces, and it even talks—if you listen.

This is also a story of people—those who lived and worked in the forest, who left their mark, and who harvested its bounty. The early explorers who first laid eyes on it, the native Indians who burned it, and the politicians who wrote the laws all played a role in this story.

The Millicoma Forest, as it was eventually named, is several hundred thousand acres of some of the richest privately owned forestland in North America. Located in Coos and Douglas Counties in southwestern Oregon, it is an industrial tree farm with a unique history. The Millicoma is remote and relatively inaccessible. The forest contains steep ridges and deep canyons and is bounded by three-hundred-foot waterfalls to the west and a precipitous escarpment to the east. Professional foresters and many forest scientists and researchers were involved in its management from the beginning of its development. The forest presented formidable engineering, silvicultural, and ecological challenges. From its dedication by the patriarchal German forester Carl Alvin Schenck in 1951 to the completion of the harvest of most of the mature timber forty years later, forest practices have run the gamut from high yield forestry to habitat conservation and ecosystem management. The forest is also a part of the economic, social, and political arena of the nation. This also must be a part of the story.

From the rolling decks of their frigates, the eighteenth-century explorers off the coast of Oregon may have been the first to see the Millicoma. If by chance they had been off the coast of southern Oregon in the summer of 1765, they might have witnessed the beginnings of a new forest—the regrowth from a fire that occurred sometime during that period. We find no evidence that any European eyes saw this holocaust, but during this period and even to modern times the coast of Oregon was truly a *costa del fuego,* a coast of fire.

Explorers from all around the world explored this area in the last quarter of the eighteenth century. The Spanish sailed off the coast of Oregon in 1774 and 1775, and then again in 1778, 1779, and 1791. The English traveled off the Oregon coast with James Cook in 1778 and with George Vancouver in 1791. The Frenchman La Pérouse passed through the area in 1786, and the Americans explored the coast during Robert Gray's voyages in 1788. Many of these explorers marveled at the thickly forested shores. Some remarked on the smoke and fires coming from those forests.[1]

The actual date of the fire that raged through the Coos River drainages is, of course, speculation—but not entirely. In 1945 and 1946, Weyerhaeuser foresters conducted an intensive examination of some 200,000 acres of Douglas-fir forests in the Coos and Millicoma basin. These foresters were probably the first to run compass lines through this still-untouched wilderness since the General Land Office surveyors and the Northern Pacific railroad cruisers at the turn of the century. Running compass lines through a forest marked by only one newly built dirt access road and trapper's trails, the foresters measured trees at regular intervals. They determined the diameter and heights and bored into the trees for the story told by their annual growth rings. Almost two hundred miles of line were run, and some fifteen hundred trees were bored. When the results were analyzed, tree after tree was close to the same age—180 years—meaning that something big had happened somewhere around 1765,

1 George Vancouver, *Voyage of Discovery of the North Pacific Ocean and Around the World* (London: 1806).

Dean Higinbotham using an increment borer in a Douglas-fir tree.

and that something big was fire. In fact, the foresters in 1945 noticed the charred scars in the scattered ancient survivors of that fire.

Almost every Douglas-fir forest in the Pacific Northwest, through the millennia, has experienced cataclysmic changes wrought by fire, windstorms, epidemics, volcanoes, and earthquakes. Of these elements, ecologists have labeled fire as the dominant factor of forest history. From studies of present forest cover, tree rings, fire scars, and carbon studies, the evidence shows Oregon forests were shaped over centuries by fire. Ample evidence exists that prior to the discovery of the New World, Native Americans continually modified forest composition.[2] In addition to the 1765 fire, which created the Millicoma Forest, the fire of 1868, which raged up the Oregon coast from Port Orford to the Umpqua River; the Nestucca fire of 1845; the Yaquina fire of 1846; and the Bandon fire of 1936 all had an impact on Oregon's forests. And the biggest of recent times, the Tillamook fire of 1933, and its subsequent reburns made headlines across the country. Over a million acres of Oregon's forests went up in smoke from the 1860s to 1970.

Reconstructing how the 1765 fire started, how it spread, and what the climatic and stand conditions were is indeed speculation—but again, not entirely specu-

2 Bob Zybach, "Native Forests of the Northwest, 1788–1856: American Indians, Cultural Fires and Wildlife Habitat." *Northwest Woodlands* (spring 1993).

lation. Climatic research scientists created a grid using densiometric data from tree rings that reconstructed summer (April through September) temperature patterns across western North America (from southern Alaska to Arizona) from 1600 to 1982. The research showed during this 383-year period, the decade of 1768 to 1777 was the warmest for the area most closely corresponding to the Millicoma Forest, although no measurements were taken from this part of Oregon.[3]

Putting together the gathered evidence, we can recreate the conditions that existed in the summer of 1765 on the Millicoma. The overmature forest, which had not burned for centuries had tons of down fuel on the forest floor that had been drying out all during the unusually hot and dry summer. East winds had turned the woods into a dry, crackling tinderbox. All it would have taken was a spark to touch it off.

Who or what touched it off? Lightning so common in the Cascades and Rockies was not that common in the Coast Range. Evidence shows that the native tribes across the country used fire to improve the hunting or clear the woods for easier travel. They burned the woods every year, but little damage was done because of the low levels of fuel. The coastal Indians, other than trekking up the coastal rivers for salmon in the fall, did not travel back into the forest. The Calapooyas in the Willamette Valley had a history of burning and might have ignited the fire. Whether the blaze was started by a torch carried by Indians or a campfire, it quickly got out of hand. From somewhere below the divide between the Coos and Umpqua Rivers, the fire roared westward, making its own wind, leaping ridges, uprooting trees, and releasing as much energy as a nuclear reaction. Two or three hundred years of forest growth went up in flames in probably not much more than twenty-four hours.

Eventually the autumn rains and winds came off the sea, and the great fire died out. A silent and desolate land stood in its wake. The fire killed trees towered over the charred landscape like a giant lathe house. In coves and patches, islands of the original forest still stood. But where did the trees that the foresters measured in 1945 come from?

The fecundity of nature at times seems boundless. The spring preceding the fire must have also been dry—perfect conditions for the dispersal of pollen and the flowering of the female flowers of the Douglas-fir. That spring, pollen fell like a golden rain, covered the trees, and lay like a film on the woodland

3 K. R. Briffa, P. D. Jones, and Schweingruber, "Tree-Density Reconstructions of Summer Temperature Patterns across Western North America since 1600." *Journal of Climate* 5 (July 1992): 735–53.

pools. Almost every tree must have had a bumper crop of cones loaded with fertile seed. In the fall after the fire, the surviving trees released their winged seeds to fall on the nitrogen-enriched, blackened soil. If the seed escaped the ravaging of mice, birds, insects, and erosion, the next spring the tiny seedlings sprang up like grass. Then began the race for survival and dominance. Along with the tree seedlings, an abundance of herbaceous vegetation poked green shoots up through the forest soil—bracken fern, fireweed, vine maple, rhododendron, all competing for light and moisture. In many cases, these species claim the burns, and years may pass before the forest becomes established. In our story however, the new forest rose like Phoenix from the ashes.

By the turn of the eighteenth century, the young forest had begun to close up. The race for light produced winners and losers. The losers died out and fell onto the forest floor, while the winners shed their lower limbs and made prodigious growth in height and diameter.

Relatively few animals lose their lives in big forest fires. They move out. Their habitat, however, changes radically. Some species thrive on the herbaceous vegetation sprouting up after a burn. For a few years after the Millicoma fire, deer and elk populations soared. Mice bred in abundance and provided food for the predators, both woolly and feathered, that hunted the burn. Woodpeckers pounded the snags for grubs. But as the crowns of the young trees merged, animal and plant diversity changed. The forest became quieter and almost impenetrable. Many decades would pass before any white man entered these remote canyons. The forest continued to grow, unseen by European eyes and untouched by humans. Wreathed in the mists and rains from the sea, caressed by the summer winds, and tossed by the winter storms, the young forest covered the once barren hills.

"I regard the question of public lands, next to that of the currency, the most dangerous and difficult of all." —John C. Calhoun, 1841.

1 / Whose Woods These Are

The Millicoma was little noticed by the world around it for almost a century after its birth. Explorers were more interested in reaping the wealth from the fur trade. The forest remained only a refuge for sometimes-hostile natives and a haven for cougars, bear, and wolves. And upon their arrival, the settlers saw the Millicoma only as something to be cleared for the planting of more crops. Who finally laid claim to these Oregon woodlands?

The Spanish had explored the area since the early 1500s. In 1513, Vasco Núñez de Balboa led his troops across the Isthmus of Panama and gazed out at the Pacific. He took possession of this vast new sea and all the lands that it washed. Spain soon held dominion over the entire west coast of North and South America—or so they thought. No one had asked the Milluks of the Coos Bay Indians, who traveled up the Millicoma River every fall to spear salmon, what they thought about the Spanish claims. They had never seen a Spaniard or for that matter any other European.

By the end of the eighteenth century, Spain's role on the west coast of North America was held to California and Mexico. A new presence on the world's stage, the United States of America, was expanding its territory. President Thomas Jefferson presided over the purchase of the Louisiana Territory from the French in 1803, and in 1804, he sent Meriwether Lewis and William Clark west to the Pacific. The U.S. acquired Florida from Spain in 1819, an event that prompted John Quincy Adams to declare, "the remainder of the continent should ultimately be ours." The British, however, still claimed the Oregon country even though the Hudson Bay Company and its Chief Factor at Fort Vancouver, John McLoughlin, held a dominant presence in the territory. Many years of negotiations with England lay ahead.

In 1823, President James Monroe announced that "the American continent by the free and independent condition which they have assumed and maintain are henceforth not to be considered as subjects for colonization by a European power." In August 1827, a convention signed by the United States and Great Britain stated that "all territories claimed by Great Britain or by the United States, west of the Rocky Mountains, are free and open to the citizens or subjects of both nations for 10 years."

In 1844, as settlers poured into the territory, negotiations between the U.S. and Britain regarding the Oregon Territory began again. As one observer noted, "It appears, however, to be certain that under all or any succeeding circumstances whether of peaceful partition of the countries in dispute or the only other alternative of war between the two claimant powers, those countries will receive their population from the United States. Nearly a thousand citizens—a number far greater than that of the first settlers in Virginia or New England—have within a few months entered Oregon and an equal number will soon follow with the determination to make it their home."

In the 1840s, "Oregon Fever" struck the country, especially in the Midwest. Prompted by politicians, the press, and promoters, many sober, God-fearing farmers sold their farms, invested in wagons, oxen, mules and horses, and headed west across the prairies, the "Great American Desert," and over the mountains to the Willamette Valley in Oregon. They dreamed of free land, even though no one yet knew who had the right to grant title. One of the most vocal of the expansionists was John L. O'Sullivan, editor of the New York *Democratic Review,* who coined the phrase "Manifest Destiny." In an editorial appearing in the December 27, 1845, issue of his publication, O'Sullivan wrote, "Away with all these cobweb tissues of rights of discovery, exploration, settlement, contiguity—our title is by the right of our manifest destiny to overspread and to possess the whole of the continent which Providence has given us."

Hall J. Kelley, who founded a society in 1829 to promote American settlement in the Oregon country, gave talks before audiences back east extolling Oregon, a place he had never seen. He finally arrived in California in 1833 and joined up with Ewing Young's great cattle drive from California to the Willamette Valley. Young, known as the master trapper, thought Kelley a monumental bore. Kelley made it to Oregon but was sick most of the way and was shocked at the brutal treatment of the Indians meted out by Young's bunch of renegades. Young made it to Oregon with 630 of the 729 cattle he started with. Accused by the Mexicans of stealing most of them, he sold the cattle to settlers and made a fortune. He died the richest man in Oregon in 1874.

Meanwhile, James Polk, wanting to take over the entire Oregon Territory, campaigned for president in 1844 with the slogan "Fifty-four Forty or Fight!" For a while it appeared that the United States could go to war with Great Britain; but instead, they chose Mexico. In 1846, the U.S. negotiated a treaty with Great Britain that set the northern border of the Oregon Territory at the 49th parallel.

The first American settlers to arrive in Oregon established a colony near Salem in the Willamette Valley in 1841. They were led by Jason Lee, a Methodist missionary from Massachusetts. Prior to establishing a permanent settlement, Lee traveled overland to Oregon with fellow New Englander Nathaniel Wyeth's second expedition. He arrived at Fort Vancouver in 1834 with his nephew Daniel, Cyrus Sheperd, and two aides. He met with John McLoughlin and set about bringing Christianity to the local natives. He moved up the Willamette Valley and was taken with the beauty of the site, which reminded him of home. He returned east and sailed to Oregon on the *Lausanne* with fifty-one settlers. Lee and his group settled some eighty miles above the Willamette River's junction with the Columbia. The hub of the settlement was a clearing on the east bank of the Willamette about eighteen miles above the Willamette Falls. The place was called Champoeg. In 1845, the settlers under the Champoeg Compact formed their own government realizing that there would be no property values without one. They provided for provisional land claims, but they needed the protection of the United States.

In 1838, Lee and later another missionary, Marcus Whitman, traveled to Washington, D.C., where Lee had Senator Lewis Linn of Missouri, an Oregon advocate, presented to Congress a petition signed by thirty-six settlers calling for U.S. government protection and laws. Thomas Jefferson Farnham, a lawyer and author of *The History of the Oregon Territory* prepared a petition to Congress signed by sixty-seven U.S. citizens. The petition set forth that the signers settled in Oregon under the belief that it was a portion of the public domain of the United States upon which they might rely for the blessings of free institutions and for armed protection. Senator Linn, almost annually, introduced bills that called for free land grants as rewards to Americans who would immigrate to Oregon. He never lived to see any of them enacted. By 1841, one estimate showed the population of what we now know as Oregon at 150 Americans. By 1846, the United States had full possession of the entire region between the 42nd and 49th parallel and from the Rocky Mountains to the Pacific. Demands by the settlers for land prompted a whole array of land laws that gave or sold the public domain lands to private hands. This went on unabated until 1892, when the first forest reserves were set aside. During the Progressive Era, Theodore Roosevelt repealed many of the most egregious land laws and removed millions of acres into forest reserves, which became the national forests in 1906.

By the 1820s, when the first settlers arrived in Oregon, the Millicoma Forest still had not been seen by a white man as far as we can determine. All of the early explorers and trappers traveled up and down the Willamette and Umpqua Valleys or along the coast. None of these early explorers had penetrated the wilderness lying between the Umpqua and the Coquille. McLoughlin's trappers found the beaver trapping poor when they ventured south of the mouth of the Umpqua in 1826. In 1828, Jedediah Smith was the first white man to bring a sizable company up the coast from California to Oregon. He met disaster when fourteen of his eighteen-man company were killed by Indians at the mouth of the Umpqua. Smith escaped to Fort Vancouver. The Wilkes expedition in 1841 produced a "Map of the Oregon Territory," which included Fort Nisqually on Puget Sound, but Coos Bay was not mentioned. The Scottish botanists, Archibald Menzies and David Douglas, for whom the Douglas-fir was named, never saw Coos Bay or most certainly the Millicoma Forest. The forest was now about eighty years old with some trees approaching two hundred feet tall and three feet in diameter. Most of the lower limbs had fallen and the bark on the clear boles were beginning to furrow. The forest floor was carpeted with sword fern and mosses. Except for an occasional windfall, the forest contained few obstacles for travel, and it was crisscrossed with elk and deer trails. Indians dug carefully covered pits to trap an unwary elk along these game trails.

Three thousand miles away in the halls of Congress, lawmakers were devising ways to encourage the settlement and development of the American West. From the very beginnings of our nation the land question was one of the most important facing the young republic. Thomas Jefferson in 1776 said, "The people who will migrate to the Westward will be a people little able to pay taxes. . . . By selling the lands to them you will disgust them and cause an avulsion of them from the common union. They will settle the lands in spite of everybody. . . . I am at the same time clear that they should be appropriated in small quantities." In 1841, John C. Calhoun stated, "I regard the question of public lands next to that of the currency, the most dangerous and difficult of all which demand the attention of the country and the government at this important juncture of our affairs." In that year the preemption laws were passed, and the event was hailed as "truly a frontier triumph." Congress determined that the national interest could best be served by transferring the public domain by gift and sale to individuals and companies that would develop the land and its resources. Also, they felt the only way to bind this vast nation together was to encourage the building of railroads. So during the years just preceding and then following the Civil War, a variety of proposals became law: the railroad land grants, the Donation Land Act, the Homestead Act, the Timber and Stone Act, the Timber Culture Act, and the Forest Lieu Act. Under the provisions of these acts, between 1850 to 1950, almost 30 million acres of Oregon's 62 million acres of public domain were transferred to individuals, companies, or the state.

By 1850, all Oregon lands belonged to the federal government except for vague, frequently ignored Indian titles. But what about all the earlier settlers who had come into the country beginning in the 1830s? Until the passage of the Donation Land Act in 1860, settlers could exercise squatter sovereignty or preemption rights to 160 acres. After the land was surveyed, a more legalistic method of title transfer was called for, hence the Donation Land Act. This act granted 320 acres to any white settler who was a U.S. citizen residing in Oregon before December 1, 1850. It required the settler to reside on the land for four years during which time he was supposed to cultivate it. If the settler was married, he received an additional 320 acres. If the settler had arrived between 1850 and 1853, he was granted 160 acres, and again it was doubled if the man were married. Women were assiduously courted by land-hungry bachelors, and one record shows a twelve-year-old girl being led to the altar, but she lived with her folks for several years after the wedding. The date for the land claims was later extended to 1855, and the requirements were further loosened when the residency requirement was shortened to two years, after which the land could be purchased for $1.25 per acre. After April 1, 1855, all public land west of the Cascades in Oregon except for Donation Land claims, mineral lands, and public reserves were subject to public sale. Jerry A. O'Callaghan in his treatise, "The Disposition of the Public Domain in Oregon," which was published by the Senate Committee on Interior and Insular Affairs in 1960, showed how Oregon's public lands were disposed:

Homestead: 11,097,982 acres
Land Sales: 6,455,551
Grant to State: 4,329,445
Donation Claims: 2,614,082
Wagon Road Grants: 2,490,890
Railroad Grants: 1,588,532
Miscellaneous: 992,921

Total: 29,569,403 acres

The forest of our story did not yet have a name, but it soon had an owner. The biggest private real estate transaction in the nation's history up to that time had its beginnings at 266 Summit Avenue in St. Paul, Minnesota. Here lived Frederick Weyerhaeuser, a German immigrant, who had made a fortune in the great white pine forests of Wisconsin and Minnesota. He was born in the Rhineland of Germany and immigrated to the United States as a young man. His first job in America was in a brewery in Erie, Pennsylvania. He disliked this job and moved to Rock Island, Illinois, where he took a job as night watchman at a sawmill. Through hard work he impressed the owners and moved up to other jobs in the mill. When the mill owners went bankrupt, he and his brother-in-law Frederick Denkman bought the mill. Soon he left the job of running the mill to Denkman and headed up the Mississippi to find timber. Throughout the woods he was known as Dutch Fred, probably because of his accent. He was a shrewd judge of both timber and men and had established the respect of many of the leading lumbermen of the region who found that they seldom lost money if they listened to Fred. He reportedly said, "I know this much: whenever I buy timber I make a profit; whenever I do not buy I miss an opportunity. I have followed this practice for many years and have not lost anything by it." He enjoyed making deals and for his time was remarkably farsighted. In St. Paul, Weyerhaeuser was blessed with a neighbor he got to know well—James J. Hill, the railroad magnate. Hill, through the immense land grants given to the Northern Pacific Railroad (NP), which was building a line from Minnesota to Puget Sound, owned immeasurably more pine trees than even Fred Weyerhaeuser and his associates had dreamed of. A good deal of Jim Hill's pine trees were Douglas-fir far out in the Pacific Northwest. Most of Weyerhaeuser's peers were more interested in the southern pine than in the wilderness on the West Coast. But that was about to change.

The first land grants to the Northern Pacific Railroad were made in 1864. Additional grants were made in 1870. The original grant specified a twenty-mile strip on each side of the completed sections of the line that passed through a state and a forty-mile strip on each side where it passed through a territory. The railroad was granted alternate sections of land within the strips. In addition, if the lands within the strips were already reserved, an additional ten-mile strip was added as indemnity lands. This was later extended by another ten miles. In essence, the railroad had a swath of alternate sections of land 120 miles wide from Minnesota to the shores of Puget Sound. By 1886, according to the General Land Office, the NP had constructed 2,021.38 miles of its road, all of which had been approved by President Abraham Lincoln. Of the road constructed, 35.80 miles were within the states of Wisconsin, Minnesota, and Oregon, and 1,669.58 miles were within the territories of Dakota, Montana, Idaho, and Washington. At this rate, in 1886 the railroad company was entitled to 47,244,288 acres and still building.

During long evening conversations with his neighbor, Weyerhaeuser learned that the railroad had to raise some money quickly for redemption of bonds that were due. To help raise the money, the railroad needed to sell some of its enormous land holdings. Weyerhaeuser and some of his associates had been to the Northwest. Weyerhaeuser's eyes must have lit up when he saw the magnificent stands of Douglas-fir that Jim Hill's people showed him. It was also apparent that the end was in sight for the white pine in the Lake States. So on January 3, 1900, Weyerhaeuser and William H. Phipps of the Northern Pacific Railroad signed the papers transferring 900,000 acres of timberland in Washington State to the Weyerhaeuser Timber Company for $5,400,000. Three million dollars of the purchase price was to be paid immediately and the remainder in eight semi-annual payments at 5 percent interest.

It took almost all of Weyerhaeuser's associates in the upper Mississippi region to raise the money. They were apprehensive about this gamble but finally trusted the judgment of Weyerhaeuser and named the newly established company after him. The news of this bold transaction spread across the country, and some in the Pacific Northwest immediately feared that the small operator would be squeezed out by the "syndicate." At six dollars an acre it proved to be a bargain indeed, but in 1900, it was a huge bet on the future.

The Millicoma Forest was hundreds of miles away from Puget Sound, so how did Weyerhaeuser end up with these lands in southwest Oregon far from the NP mainline? In 1899, Congress established Rainier National Park after some years of intensive lobbying by Tacoma and Seattle boosters for Congress to protect this beautiful mountain. Officials within the Department of Interior had advised Congress not to establish the park until the railroad claims had been exchanged for other federal lands. The NP lands that were taken up in the park could, under the provisions of the Lieu Land Act, be exchanged for other federal lands. The railroad issue stalled the Washington National Park bill long enough to prevent its passage in 1897. Supporters tried again in 1898 and 1899. By that time the bill had been rewritten to grant the NP lieu land rights in exchange for its lands on Mount Rainier. The railroad had built relatively few miles in Oregon. Charges were made that the NP had framed the bill that established the park and openly lobbied for the passage of the Enabling Act of March 2, 1899, which created the nation's fifth national park. Critics claimed that under the provisions of the Lieu Land Act the NP was exchanging glaciers for rich timberland. According to some figures, the Mount Rainier Forest Reserve deal netted the railroad 540,000 acres, of which 320,000 acres were Douglas-fir in Oregon, 100,000 acres of fir in Washington, and 120,000 acres of pine in Idaho.

The Progressive Era of Teddy Roosevelt witnessed an outcry against the widespread land frauds in the West. Federal investigations resulted in the conviction of a United States senator from Oregon, and many public officials went to jail. The so-called muckraking journalists began writing stories about the mysterious Weyerhaeuser whom they claimed was richer than Rockefeller. One article described him as an "octopus." The always shy and now aging Weyerhaeuser was dismayed and stoutly maintained that every dollar he had made was earned honestly, and no evidence indicated otherwise. However, C. A. Smith of the Coos Bay Lumber Company and R. A. Booth of Booth-Kelley Lumber Company were both charged with using illegal dummy entrymen to secure valuable timberlands. George Long, Weyerhaeuser's manager in the West would not condone this practice. The Weyerhaeuser's close buyer-seller relationship with the NP continued, however, and in July of 1902, Weyerhaeuser purchased 31,000 acres of NP land in Oregon for about five dollars per acre. It was the first of many acquisitions on the Millicoma.

After four centuries of conflicting claims, much of this remote forested area became the property of Weyerhaeuser's timber company. Neither the Spanish nor the British had ever seen this forest, nor had the Weyerhaeusers. What had they purchased?

David Douglas, a young botanist who traveled through Oregon in 1826 sponsored by the Royal Horticultural Society of London, described the Douglas-fir, which now bears his name, as some of the most striking and graceful objects in nature. The trees in the Millicoma in 1902 were now clear of limbs for one hundred feet or more. The trees in the creek bottoms were so much taller than those on the ridges so that if the forest could have been viewed from the air, the terrain would have appeared much more gentle than it really was. The understory contained hemlock and cedar. The forest floor was carpeted with sword fern, mosses, oxalis, and trillium. Devil's club and salmonberry grew along the creeks. The openings left by windfalls were commonly choked with tangles of vine maple and salal.

Elk foraged throughout the forest. Red squirrels chattered in the tops. At dusk the wide-eyed flying squirrels would drift down through the trees. Marten preyed on the squirrels, cougar on the elk, and bark beetles on the fir. With each year the trees added another wide ring of wood. To lumbermen this was still red fir, not to be valued as highly as the much older yellow fir with its dense, close-grained, clear wood. But it was indeed a treasure chest and growing more valuable every year.

References

Bancroft, Hubert Howe. *History of Oregon*. San Franciso, Calif.: The History Company, 1890.

Collins, Dean. "Ewing Young's Cattle." *Oregon Journal* (1933).

Dodge, Orvil. *Pioneer History of Coos and Curry County*. Salem, Ore.: Capital Printing Company, 1898.

Douglas, David. *Journal of David Douglas*. London: W. Wesley, 1914.

Farnham, Thomas Jefferson. *History of Oregon Territory: It Being a Demonstration of the Title of These United States of North America to the Same*. New York: W. Taylor, 1845.

Fremont, John Charles. *The Expeditions of John Charles Fremont*. Edited by Donald Jackson and Mary Spencer. Urbana: University of Illinois Press, 1970.

Gates, Paul Wallace. *Public Land Policies, Management and Disposal*. New York: Arno Press, 1979.

Gillette, Jane Brown. "240 Summit." *Historic Preservation* 46, no. 4 (July/August 1994).

Greenhow, Robert. *The History of Oregon and California*. Los Angeles, Calif.: Sherwin and Freuntal, 1970.

Hidy, Ralph, Frank Hill, and Allan Nevins. *Timber and Men: The Weyerhaeuser Story*. New York: Macmillan Company, 1963.

Holmes, Kenneth L. *Ewing Young, Master Trapper*. Portland, Ore.: Binford and Mort, 1967.

Lamar, Howard P., ed. *The New Encyclopedia of the American West*. New Haven, Conn.: Yale University Press, 1998.

Martinson, A. D. *Wilderness Above the Sound*. Flagstaff, Ariz.: Northland Press, 1986.

O'Callaghan, Jerry A. *The Disposition of Public Land in Oregon*. New York: Arno Press, 1960 [reprint].

Pethick, Derrick. *First Approaches to the Northwest Coast.* Vancouver, B.C.: Douglas and McIntyre, 1976.

Puter, S. A. D. *Looters of the Public Domain.* New York: DeCapo Press, 1972 [reprint].

Robbins, Roy Marvin. *Preemption: A Frontier Triumph.* Cedar Rapids, Iowa: The Torch Press, 1931.

Robbins, Roy Marvin. *Our Landed Heritage: The Public Domain, 1776–1936.* New York: P. Smith, 1950.

Salo, Sarah Jenkins. *Timber Concentration in the Pacific Northwest.* Ann Arbor, Mich.: Edwards Brothers, 1945.

Sparks. W. *Report of the Commissioner of the General Land Office Relative to the Lands Granted to the NP Railroad Company.* U.S. General Land Office, April 20, 1886. Washington, D.C.: Government Printing Office, 1886.

Twining, Charles E. *George S. Long, Timber Statesman.* Seattle: University of Washington Press, 1994.

"We ought to get as young a class of men as possible who have all the desirable qualities we need."
—George Long

2 / What Do We Have Here?: Explorers, Trappers, Surveyors, and Timber Cruisers

Coos Bay, the only deep-water harbor other than Humboldt Bay between San Francisco and the Columbia River, was little noticed by the early seafarers. David Douglas, the botanist, set off with some of the Hudson Bay Company's trappers in the fall of 1826 to explore and collect specimens from the Willamette Valley and the Umpqua. He complained in his journal about the rain, which fell in "torrents" most of the time. Although Douglas did not go all the way down the Umpqua, he wrote in his journal of a report made to him by the trapper Alexander McLeod, who might have been the first white man to enter Coos Bay. Douglas, describing McLeod's journey from the mouth of the Umpqua, wrote, "He journeyed along the sea-beach for about 23 miles when he came to a second river, similar in size to this one also affording the same sort of salmon and salmon trout. At its mouth were numerous bays and in one of the said bays he pursued his route to the south in a canoe for twenty miles where he came to a third river, a little smaller than the others, which by the Indian accounts takes its waters a long distance in the interior. (the Coquille?) According to the Indians a large stream falls into the sea about 60 miles further south." (the Rogue?)

The next time we hear anything about Coos Bay is when the ship *Captain Lincoln* went aground on the North Spit beach on New Year's Day, 1852. The ship carried a company of Dragoons who built a winter camp from the remains of their ship. For four miserable months they lived in their tent camp made from the sails of their ship. Indians and the few settlers in the area visited "Camp Cast-Away" to trade and gawk at the soldiers. While awaiting rescue the castaways discovered the large bay behind them with many sloughs and inlets. They eventually made it down the beach to Port Orford.

Settlers were arriving, and the first Donation Land claim on the southern Oregon coast was filed by Thompson Lowe at Bandon. John Yoakum settled at Empire in 1853 and John Riley Davis was the first white man to build a cabin on the Coos River. The Coos Bay Company arrived at Empire in 1853, and the Baltimore Colony settled at Myrtle Point in 1859. The town of Marshfield was settled in 1853, and Asa Simpson built a sawmill and shipyard at North Bend in 1856. The first ship went down the ways in 1857, and the first lumber went out in 1858.

The wilderness beyond the bay was still unknown and untouched by Euroamericans. We can only speculate as to who was the first non-Indian to set foot on our Millicoma Forest. He or she may have been guided by an Indian, or perhaps it was a teenager who told his mother he was going to go up the creek to explore.

The settlers continued to come by land and by sea. One wagon train that left Indiana in 1853 included an eleven-year-old boy traveling with his parents. That boy was destined to become one of the most important and respected citizens of Coos County. Some twenty years later, Simon Bolivar Cathcart settled on the South Fork of the Coos River. In 1874, he was elected county surveyor of Coos County, a position he held for the next twenty-six years. As surveyor he was a very important man indeed. For land to be claimed or sold it had to be surveyed and platted by township, range, and section—no easy task in the rugged coast range. The history of the rectangular survey system is told in Gilbert White's 1983 monumental work for the Department of Interior.

The rectangular system of surveying public lands goes back to Thomas Jefferson, who was deeply concerned about the lands west of the original colonies. Jefferson wanted the small farmer to have access to these lands rather than large landholders or land companies. In 1785, the Continental Congress passed the Land Ordinance, which established a system of subdividing lands. It ordered that lines should be measured with a chain and plainly marked with "chaps" on the trees and exactly described on a plat. The surveyor was, "to note as to their proper distance all mines, saltsprings, saltlicks and mill seats that shall come to his knowledge; and all water-courses, mountains and other remarkable and permanent things over and near which such lines shall pass, and also the quality of the lands." He was also required "to pay the utmost attention to the variation of the magnetic needle and shall run and note all lines by the true meridian certifying with every plat which was the variation." And for these services the surveyor would be paid at the rate of two dollars for each mile run. This would include the wages of chain carriers, markers, and any other expenses incurred. By 1804, the rate had risen to four dollars per mile.

Years later, on September 27, 1850, Congress passed an act that created the Office of Surveyor General in Oregon and extended the rectangular system to the territory. In October of that year, William Gooding, perhaps hearing about the steepness of much of the Oregon country, respectfully declined the position. In November, John B. Preston of Chicago was appointed. After a briefing in Washington, he departed for Oregon with four solar compasses, transit, sextant, and chains. He went by sea to Panama, crossed the Isthmus and continued by ship, arriving in Oregon City in May, where he set up an office. Remembering the burning of the city by the British in 1814, the people back in Washington cautioned him about keeping his records safe from fire. The commissioner of the General Land Office (GLO) assigned his clerks and any other help he could round up to remove all the records and hide them in private homes. The records survived, but the GLO burned to the ground. Preston established the initial point for the Willamette Meridian just west of Portland and by the end of May had let contracts for the first surveys. The meridian line running south was surveyed by James E. Freeman from Wisconsin, and the baseline west to the coast range and east to the Cascades was run by a Michigan man, William Ives.

By 1853, the pay for surveying had risen to fifteen dollars per mile. The surveyor general complained that it was difficult to get any qualified surveyors at that rate, which was set by law. The government soon found that they were getting what they paid for. One surveyor admitted that it might be necessary to "sacrifice scientific accuracy for speed." In one case the original surveyor had set off the magnetic declination at 9 degrees West instead of 9 degrees East, which meant all the lines were off by 17 or 18 degrees. The shoddy work of

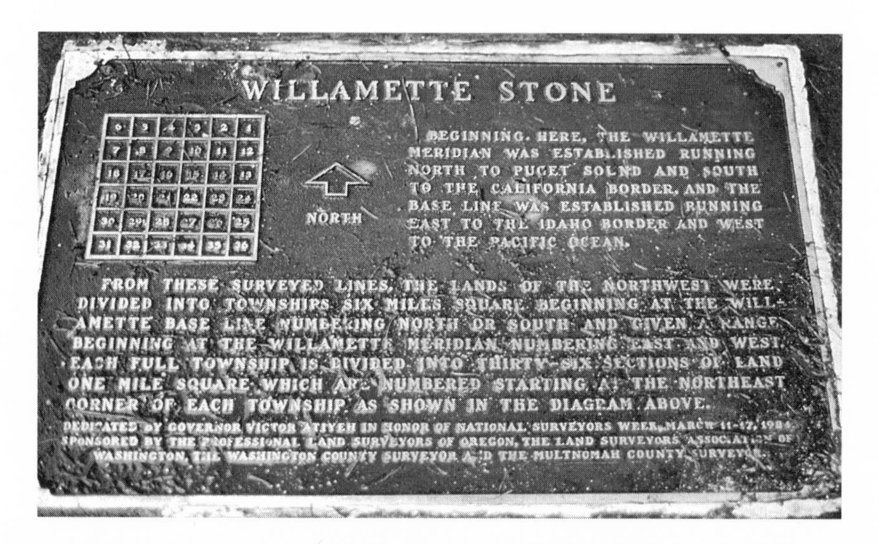

The Willamette Stone in Portland, Oregon, marking the Willamette Meridian, from which land surveying began in the Pacific Northwest. Photo Arthur V. Smyth.

some surveyors resulted in a need for many resurveys. In 1851, the Washington office published the *Oregon Manual of Surveying Instructions,* which was revised in 1855 and remains to this day essentially unchanged.

The Oregon surveyors were running into problems with Donation Land claims, which frequently ran in all directions with little regard for any subdivision lines. As the claims were filed and verified the Donation Land claim boundaries had to be surveyed and tied into the GLO grid. One claim crossed the Umpqua River even though the bed of the river was patented. This must have resulted in a field day for lawyers. The deputy surveyors were required to take an oath that stated, "I will well and faithfully and to the best of my skill and ability and according to the laws of the United States and the instructions of the Surveyor General perform the duties so confided to me as I shall answer to God at the great day." One California surveyor must have had a lot to answer for on his "great day." He and his gang were convicted of a gigantic swindle wherein all the surveying was done in a "boiler room" in San Francisco. It was only when the settlers failed to find any corners that the plot was uncovered.

These surveyors, with their chainmen, axemen, and compassmen, were among the very first non-Indians to set eyes on the Millicoma Forest. As they blazed the trees, brushed out the line, called out the tallies, and set the corners, they must have marveled at these towering trees. One of the first surveys was that of portions of Township 24S Range 9W, which is drained by Lake Creek.

W. H. Byars, under Contract 292, began his survey on August 7, 1878, and completed the job on September 16 in the same year. He mapped and surveyed 10,516 acres and ran twenty-nine miles of line subdividing the eastern and northern portions of the township. All he had to say was, "the lands along the creek are first class generally covered with fine maple and other timber. These lands when cleared will be very valuable for agricultural purposes. The hills are generally covered with Fir and its undergrowth." In 1878, timber was still regarded as an impediment to growing crops.

We do not know how Byars got into this country well over a century ago, but he and his crew probably packed in on a trail from Glenn Creek to the west. This was the route followed by homesteaders who came into Lake Creek some thirty years later and also the route of the first access road built by Weyerhaeuser in 1945.

Some additional surveying was done by Oscar Theil in 1898. The remainder of the township was surveyed in 1905 by Levi C. Walker. By now the surveyors found evidence of homesteaders back in this rugged wilderness. The plat approved by the surveyor general on July 9, 1897, of T24S R10W showed a house on Matson Creek, which marked the Dan Mattson homestead. On Glenn Creek the plat showed the Joe Schapers and the Hanson places. Around 1906, the Guerin family from Coquille took up homesteads on Lake Creek and Bear Camp Creek in T24S R9W. Cabins were built, some timber was felled, and orchards were planted. On the high ridge between the East Fork of the Millicoma and the South Fork of the Coos, claims were laid out by Cook, Cortes, and King. Other than the Glenn Creek homesteaders, and possibly the Lake Creekers, most of these people were more interested in timber claims than in settling the country.

The remarkable story of the families who packed in above the Golden and Silver Falls and hewed out their homesteads in the narrow valley of Glenn Creek is told in Lionel Youst's book *Above the Falls: An Oral and Folk History of Upper Glenn Creek, Coos County, Oregon*. Youst has captured the lives of a breed of people who no longer exist. Their counterparts could be found in the "hollers" of Appalachia, the bayous of the South, the sodhouses of the prairies, and the mountain men and trappers of the Rockies. They lived off the land; they built the schools for their children; and they built the trails and the roads that eventually ended their isolation.

Beginning in the 1880s up to the middle 1950s, these pioneers lived in a little-known corner of what was a forested wilderness. Now that they are gone, the forest is once more reclaiming the land, erasing the meadows, covering the orchards and the fences. Unlike the people, primarily dairy farmers, settling

Golden Falls. Photo Arthur V. Smyth.

on the tidewater farms below them, the families above the falls lived in and on the forest. They shot elk and deer in and out of season; they trapped marten, mink, coon, cougar, and bobcat; they picked ferns and peeled cascara bark; they made moonshine; and when they needed money they worked as loggers and sawed lumber.

These were resourceful and independent men and women. They could move mountains, which in effect they literally did. The only access to this remote area for the first settlers was by boat to Allegany at the head of tidewater on the North Fork of the Coos and then a trail or sled road some ten miles to Glenn Creek. There they were confronted with what an early visitor described as "an amphitheater of solid rock." Over the rock plunged a falls that was named Golden Falls for Doctor Golden, a member of the party. The falls made a spectacular sight as they thundered into the canyon. Just to the west another falls, Silver Falls, cascaded down this amphitheater of rock. Getting around the falls seemed like an insurmountable obstacle, but these early pioneers managed it. Until 1901, nothing but a pack trail over the falls existed, but in that year "three crazy Swedes" blasted a trail over the falls, and a few years later it was widened to a road. Charlotte Mahaffy, in her book entitled *Coos River Echoes,* recounts the story of how these Swedes got through this impassable bluff. Joe Larson, a Glenn Creek settler, had experience as a powder "monkey." He talked his fellow Swedes, Alfred Tyberg and John Hendrickson, along with George Schapers, to help him. During fifty long days during the summer of 1901, they blasted out a trail around the falls. Each morning each man packed in fifty pounds of dynamite on his back. Then they would let themselves down the cliff by ropes and hand drill six-foot holes, which they would load with dynamite. At noon they would set off the charges. Then they went back for more dynamite, drilled more holes, and by the end of the day they would set off these charges. By the end of the summer they completed the trail. By 1912, advertisements announced trips to the Umpqua from Allegany through "some of the most scenic country in the West." If one left Marshfield at 5:00 A.M. the boat would get you to Allegany by 8:00 A.M., and if all went well you could make it to Drain on the other side of the Coast Range by the end of the day.

The families in this remote valley experienced marriages, births, deaths, and all the ups and downs of any family. During the three quarters of a century that humans had inhabited the upper Glenn Creek valley, the nation had seen two presidents assassinated, two world wars, a catastrophic economic collapse, and technological changes that the "three crazy Swedes" hammering their way through the rock could not have imagined. Through all this change the great forest stretching out to the east of their little homesteads remained, seemingly unchanged.

In 1941, the greatest war the world has ever seen changed the nation forever. There were changes in our forest too. For one thing it was growing, but there was death too. Throughout the forest, windthrown trees lay across the slopes with their huge roots and torn up earth thrusting through the brush. Beneath the bark of the windthrown firs the tiny Douglas-fir bark beetles were laid their eggs. In the spring the beetles would emerge and attack standing trees. Some trees would pitch out the intruders, but many would die. Patches of snags scattered through the forest and openings were filled with hemlock seedlings and vine maple. But overall this was still a healthy, vibrant forest; still adding one growth ring after another, still reaching for the sky.

After the surveyors came the cruisers—woodsmen skilled in placing value on forestlands. Probably the first of these to estimate the still untapped wealth of the Millicoma Forest were the Northern Pacific cruisers. In 1900, cruisers Peterson and Robertson worked on the NP scrip selection lands in T25S R9W in the Fall Creek drainage. Access into this country was by trail up Fall Creek and Bottom Creek from the south fork of the Coos River or by trail up the East Fork of the Millicoma River. Throughout the township these early cruisers ran across the abandoned ruins of homesteader cabins. Some were made of logs and some were framed with split boards. A few even had windows. One little clearing in the deep forest had a few berry bushes and fruit trees. The cruisers estimated that most of the claims had been abandoned in the last five to ten years. What dreams these pioneers must have had, packing into this remote wilderness, locating the corners, and laying out their claims. But their dreams were crushed by the overwhelming forest.

The county cruised the same sections in 1928, and the assessor valued the old growth fir at twenty-five cents per thousand board feet and the "white" cedar at $1.30. The first Weyerhaeuser cruises were made by A. W. Fenton from 1924 to 1928. Clarence Ross made cruises in 1929. The early cruises estimated the volume at thirty-seven thousand board feet per acre, far below what eventually came off these lands.

Timber cruising in the rough inaccessible rain forest of the Pacific Northwest was no walk in the park. Steep slopes, slippery windfalls, impenetrable brush, winter rains, and summer bugs, all called for a remarkable breed of man. Millions of dollars were riding on the skill and integrity of the cruisers. George Long, Weyerhaeuser's first manager, sent out to Tacoma after the initial purchase from the Northern Pacific, was somewhat skeptical of some of the NP's cruises and was eager to get his own cruisers out on the land. He soon realized that the cruisers he was bringing out from Wisconsin were to face conditions far different from the park-like pine forests of the Lake States. Long wrote to an associate in Wisconsin, "It seems that a man who can go into the woods out here, and do work at a rate that is satisfactory has to be in tip-top physical trim; and it occurs to me, that as inasmuch as we propose to bring some men from Wisconsin, we ought to get as young a class of men as possible, who have all the other desirable qualities we need." According to one story, years later a member of the Weyerhaeuser family stood on a roadside with the company's head cruiser, Ted Gilbert, looking down into a seemingly bottomless canyon. Weyerhaeuser said to Gilbert, "You mean to tell me you go down into that hole?" Gilbert replied, "Yes, sir, and out the other side too," to which Weyerhaeuser said, "I wouldn't go down there for all the money Weyerhaeuser has."

In 1946, the Weyerhaeuser Company's forestry department began the Coos Bay Growth and Yield Study. This was the first time in the company's operations that professional foresters made an intensive study of a forest holding before any logging took place and before any sawmill had been built. The company had owned much of the land since the turn of the century, yet the cruise information was sketchy at best. The Pillsbury tract on the upper reaches of the Coos drainage was acquired from the Pillsbury family of Minneapolis in the 1940s and had not been burned in the 1765 fire. This part of the Millicoma was much older with trees up to three hundred years old. One of the objectives of the study was to compare the net growth of the older stands in the Pillsbury tract with that of the Millicoma area tributary to Coos Bay. The decision had already been made, however, to build a mill on the bay.

The boss of the crew that had been assembled to begin the Growth and Yield Study was a cruiser, albeit he was also a graduate of a forestry school. Inners Herrala was a solidly built, blue-eyed, blond, hard drinking Finn who first introduced me to the road over Golden and Silver Falls. He had picked me up in Coos Bay, and we were headed for our tent camp at the end of the Strong-McDonald access road, which had just been pushed through from the Wilkenson place above the falls to the Lake Creek divide. After catching the Enegren ferry across the Coos he kept telling me, "Wait till you see this road." After tearing along the gravel road to Allegany I could hardly wait! In the dark of night we crept across the narrow road chiseled out of the side of the canyon. He stopped the pickup and said, "You have got to see this." All I could see was a bottomless canyon filled with the roar of the falls. But later that week I saw the falls in the daylight, and I thanked my guardian angel.

Our task that summer was to determine the site quality of the land, age of the timber, rate of growth over the last fifty years, mortality, species composition, defect in the timber, and volume. Two-man crews ran lines one mile apart across drainages measuring one-quarter-acre plots at ten chain intervals (660 feet). On each plot we tallied all live trees by diameter and species. To determine growth over the past fifty years we used an increment borer to remove a core of wood. We recorded standing dead trees by diameter and the estimated years dead. We noted conks and rot. By the end of the project we measured a total of 1,576 plots and bored 1,466 trees. We now knew a lot about our forest.

The study was highly productive. The Millicoma had a higher percentage of Site I land than much of Weyerhaeuser's properties in the Pacific Northwest. Ninety percent of the stand was made up of Douglas-fir, and almost half of the forest was 180 years old. Not counting the younger timber, mostly west of Allegany, the stand totaled over 5 billion board feet. The 180-year-old stands were producing at a growth rate of 282 board feet per acre per year or 25 million board feet total. This growth rate was reduced principally by the losses from the Douglas-fir bark beetle to a net growth of 38 board feet per year. Defect amounted to 6 percent in the 180-year-old stands, but the older stands in the Pillsbury tract had 40 percent defect and were losing volume at the rate of 40 board feet per year. The trees averaged over three feet in diameter, and one tree in the Matson Creek bottom measured 330 feet in height.

The inventory crew that amassed all this data worked out of the former Strong-McDonald camp, named for the construction company that had built the first road into the forest. It was nothing more than a dirt road. The camp was a primitive affair indeed. It consisted of an open pit latrine, water from a spring, and wooden floored wall tents for the eight to ten young men who daily pushed their lines of sample plots deeper into the forest. We had a mule and a dump truck we named Dumbo. I don't know how we wound up with a dump truck, but just after the war, vehicles were still in short supply. Dumbo would be used to transport the crews to their jumping off spots. We also had a cook, Adolph Heyer. He was brought down from Washington where he had worked in logging camps and restaurants on the skid roads of Seattle and Tacoma. Ruddy-faced and portly, one of his favorite expressions was "Fifty-two years for hire." Like so many camp cooks he was used to being away for long periods from the city lights, but once he got to town he tried to drink the place dry. We found this out when Adolph would go to town with the group detailed to buy provisions for the camp. He would slip away and in an incredibly short time down enormous quantities of beer. He usually managed to smuggle a bottle back to camp, which meant that for a day at least our cook would be flat on his stomach on the floor of the cook tent. His first comment on awakening was, "For Christ's sake someone give me a blast!"

When sober, Adolph was a delight, full of stories and quite a good cook. His morning hotcakes cooked on the huge wood-burning stove were a treat, that is if you ignored the dirty piece of burlap he used to grease the top of the stove. He claimed he was famous in Seattle for his champagne-flavored oxtail stew. He had a running battle with the mule who would sneak into his stores and eat the potatoes. Adolph was afraid of the mule and would yell at me, "Smitty get this Goddamned mule out of my spuds!"

As the lines were pushed deeper into the forest we spent more time getting back to camp than we did sampling. At this point, we would pack all the food two of us could carry on packboards. Then for about five days we would camp out in the woods or in trapper cabins that we ran across. This was no picnic in the woods. The days were long, the country rough and tough, and we worked a month straight through and then took our time off by going home to Tacoma where we all lived. But it was an adventure. All of us were somewhat awed by what we were seeing, and the fact that we were the first people into much of this untouched forest was not lost on any of us.

So after some twenty-eight days of camping out we would hit Coos Bay for the train ride home for our break. We were tough, or at least we thought we were, and we had seen country that few people had ever seen. There were no Adolph Heyers among us, but we did let off steam. After cleaning up in a hotel and soaking in a hot bath, we would descend upon Jack Ripper's Cafe and Bar in North Bend where we would eat and drink until train time around 6:00 P.M. We called the Southern Pacific passenger train, "The Coos Bay Rocket," which it most certainly was not. The train would rattle through the night over innumerable trestles to Eugene. There our car would be placed on a siding, and at 3:00 A.M. we would be coupled to the northbound train that had come over the Cascades through Oakridge. By breakfast we would be in Portland, and by noon we were home in Tacoma.

Weyerhaeuser had, over the years since the first purchase from the Northern Pacific, continued to block up their ownership by trades with the government and purchases from other owners. In the 1940s the company traded lands of equal value with the Bureau of Land Management, which helped block up both ownerships. In 1944 the company purchased over 46,000 acres of old growth timberlands from the Pillsbury family of Minneapolis for some eight million dollars. The Pillsbury tract was on the eastern edges of the Millicoma. Several townships were now almost entirely company ownership. No other Weyerhaeuser property in the Northwest was as consolidated as the Millicoma.

Howard Henderson, the company's land agent at Coos Bay for many years, was responsible for putting together many of the small, scattered ownerships within the boundaries of the Millicoma. Henderson was a short, wiry, bandy rooster woodsman, somewhat suspicious of all these college boys who had suddenly invaded his turf. He was interviewed by Lionel Youst in 1992, when Howard was in his nineties. He commented to Youst that he had experiences "that would stop a clock." Indeed he had. One escapade was the night he and a few other drunken cruisers smuggled a live goat into Ted Gilbert's room at the Umpqua Hotel in Roseburg. Henderson also gained some fame when he knocked out a state senator after an argument at the Blue Moon Tavern in Coos Bay. The Senator towered over Henderson, but he never knew what hit him.

The stage was now set for tapping the wealth of this 200,000-acre block of untouched forest. For over half a century the company had held this forest. Now it was time to bring it under management. It would take a good deal more than "three crazy Swedes" with hand drills and dynamite to bring logs out of this country.

References

Dodge, Orvil. *Pioneer History of Coos and Curry Counties.* Salem, Ore.: Capital Printing Co., 1898.

Douglas, David. *The Journal of David Douglas.* London: W. Wesley, 1914.

General Land Office Plats. Township 24S R9W and Township 24S R10W.

Mahaffy, Charlotte L. *Coos River Echoes.* Portland, Ore.: Interstate Press, 1965.

Peterson, Emil R. *A Century of Coos and Curry.* Portland, Ore.: Binford and Motts, 1952.

Puter, Stephen A. Douglas. *Looters of the Public Domain.* New York: DaCapo Press, 1972 [1857].

Twining, Charles E. *George S. Long, Timber Statesman.* Seattle: University of Washington Press, 1994.

Weyerhaeuser Company. *Coos Bay Growth and Yield Study.* Tacoma, Wash.: Weyerhaeuser Timber Company, 1947.

White, C. Gilbert. *A History of the Rectangular Survey System.* Washington, D.C.: U.S. Department of Interior, Government Printing Office, 1983.

Youst, Lionel. *Above the Falls: An Oral and Folk History of Upper Glenn Creek, Coos County Oregon.* Coos Bay, Ore.: South Coast Printing, 1992.

"Coos county is a strange and different country."

—R. P. Conklin

3 / Harvesting the Forest: The Early Planning

The economy of the Coos Bay region had, from the very beginning, relied on logs, lumbering, and shipping. And since it depended on a commodity market subject to cycles driven by interest rates, the community experienced boom and bust times. The years after the war saw eleven million young men and women returning to civilian life, eager to start families and buy homes, and they brought boom times to Coos Bay. There were a few large mills, countless small mills, and hundreds of small logging concerns. Many of these operations were involved with heavy capital outlays for logging equipment and trucks, and yet many more were mom and pop outfits with mom frequently doing the bookkeeping and sometimes driving the truck.

The war was still going on when Weyerhaeuser came to town. Although Weyerhaeuser had paid taxes and fire patrol assessments for decades, no Weyerhaeuser employees lived in the community. The very first was Robert P. Conklin, a logging engineer, who had made his name at the company's huge operations at Longview, Washington, on the Columbia River. He served briefly in Oregon at Molalla, where he helped clear logs out for the war effort. In November 1943, with the country still at war, Charles Ingram, the company's general manager at Tacoma, sent Conklin to North Bend, Oregon, to become the first manager at what was then seen as a remote outpost in the Weyerhaeuser domain. Conklin was a logger, but he was no Paul Bunyan. This trim, handsome, silver-haired engineer was a far cry from the snoose chewing, rough-talking, hard-drinking, bull-of-the-woods type that pushed the face of the timber farther and farther back into the mountains. He was a deeply religious man, married with two children, but he enjoyed the camaraderie and hi-jinks of his logger peers, and they appreciated and respected him. It must have been a dream job for a logging engineer. An untouched forest with formidable engineering obstacles, and it was his to map and plan for the orderly, long-range development and the spending of untold millions of dollars of his employer's capital.

In one of his early reports to Ingram, Conklin wrote, "Briefly it is observed that the timber is mainly Red and Bastard fir which will produce a high percentage of structural and common grades. Very few clears can be expected from the areas I have seen so far. The ground is steep and the ridges broken which indicates truck rather than railroad. Rock, a sandstone formation, lies close to the surface." On a personal matter he wrote on November 10, 1943, "There are no houses to be rented or purchased in this area. Naval officers are advertising for places in the local paper offering rewards for information leading to rental or purchase of homes." But in January 1944, Conklin moved out of the Coos Bay Hotel, purchased a home, and moved his family to North Bend.

On February 10, 1944, the *Coos Bay Harbor*'s headline read, "Weyerhaeuser To Enter Coos Lumber Production Field: Owns Vast Forests Throughout County and Pacific Coast." It also noted, "Mr. and Mrs. Conklin have taken a home in Simpson Heights and Mr. Conklin has opened a temporary office in the Odd Fellows Building." The headlines in the February 1 edition of the *Coos Bay Times* announced, "Weyerhaeuser Buys Coos Bay Logging Company," next to the headline, "German Thrust Peril Allies in Italy." Conklin was asked by the *Times* reporter as to when Weyerhaeuser would begin operations. He replied, "When, where, how much we don't know but our hope is to foresee continuous operation and sustained production from the start."

In 1944, Bob Conklin spent little time in his office in the Odd Fellows building. He was in the field looking for passes through the ridges, looking for rock, and thinking of logging railroads and truck roads. He carried on an extensive correspondence with Minot Davis, the timber and land manager at

(left) Charles Ingram, Weyerhaeuser General Manager. Courtesy Weyerhaeuser Archives.

(right) Minot Davis, Forest Engineer. Courtesy Weyerhaeuser Archives.

corporate headquarters in Tacoma, Washington. Davis was convinced that the country could be logged by railroad and for that he needed good topographic maps. He told Conklin in April of 1944, "There seems little doubt that we will have to build a first-class railroad even though this necessitates a number of expensive tunnels and expensive construction in general." Conklin pointed out, "Coos County is a different and strange country compared with the Oregon Cascade Mountain tracts and the Washington operations. It is evident that logging, transportation, forestry, and manufacturing plans will not fit a general pattern." He complained that he had only two survey crews in the field running control for topographic mapping. He wrote Davis, "The men are inexperienced but willing, two are under 18 years old and the remainder are 4Fs and one has only one arm." Both Davis and Conklin agreed, at least in 1944, "that aerial maps are poorly fitted for use in planning modern logging." Del Hilliard was the chief of party running the survey crew and remarked on the brushy understory. "If it hadn't been for the elk trails we would never have gotten through the country. They made good trails." Rex Allison, who later became woods manager, worked one summer as an instrument man for

Hilliard. He remarked, "It was very tedious work. Mules were assigned to us, but after a week they would run away for home with all our food."

In June 1944, Tacoma sent Don McKeever down to look over this "strange and different country." McKeever, a Forest Service scientist, was hired by Dave Weyerhaeuser in 1941 to establish a forest research department. His first impressions of the Millicoma were discouraging. After a three-day reconnaissance he reported, "The universal steepness of the topography, the underlying sandstone covered with a thin layer of soil, the hardness and relatively short life of the timber before maturity, the competition of woody shrubs to young trees, and the urge of production men to give the forestry aspects little consideration are all discouraging factors to a forester." In a letter to Charlie Ingram on June 20, Conklin wrote, "I am wondering if Don isn't a bit pessimistic about some of the prospects of leaving the ground in a favorable condition to promptly grow a second crop." He went on to say, "perhaps Don will feel better when he becomes better acquainted with the problems here. To me, however, these problems present a challenge. Someone is going to extract the logs, manufacture them and manage the lands in such a way that the new crop will grow. This need not be a liquidation operation . . . a forester can be the logging superintendent's right hand man in the planning of the logging shows." After the results of the Coos Bay Growth and Yield Study were compiled in 1946, McKeever did indeed change his mind. While the ridge tops and southwest slopes may have had thin soils, the bottoms, benches, and coves had some of the most productive forest soils in Oregon.

All during 1944, Conklin and Minot Davis pondered the questions of how to develop this challenging forest property. Go up the South Fork of the Coos first to tap the old growth on the east side, or up the Millicoma to harvest the red fir? Build railroads or truck roads? Where can we get rock? In a long letter, Conklin shared his thinking with Davis: "You will agree, I am sure, on one basic conclusion, that is, that our logging area is divided into two natural units which will require separate transportation systems. The Millicoma-Lake Creek area is one and the South Coos area is the other. Certainly it would be unsound to plan for the simultaneous development of both of these divisions . . . our reconnaissance surveys to date point toward a full truck transportation development of the Millicoma-Lake Creek section and a truck fed railroad tapping the South Coos area . . . an element favoring the initial development of the South Coos tract is that the more extensive stands of old growth defective timber are in this area. From a forest management standpoint these stands should be the first to be liquidated. I am inclined to suggest that a railroad be built the full distance from the terminal at tidewater to the South Pillsbury tract and that logging be started at the rear end." Conklin again lamented that, "Referring again to the northern area of our timber we have yet to discover a source of hard rock suitable for road surfacing."

A short time later, Conklin had changed his mind about railroad logging. He wrote again to Davis and said, "As to the South Fork of Coos River, I am rapidly reaching the conclusion that this will very possibly call for a truck road instead of a railroad . . . I shall have some figures from some of our operations which show in a rather startling manner that the cost per Mft/mile of railroad transportation is rapidly approaching that of truck transportation . . . there appears to be a good deal of evidence that railroad transportation labor has lost much of its efficiency . . . It seems quite possible that the cost of building a really economical railroad up the South Fork would be fully double the cost of building a first class truck road."

As 1944 was winding down, the nation was still deep into the biggest war the world had ever seen. But the country was confident that it would not be much longer before their boys would be coming home. In October, Weyerhaeuser's president, J. P. "Phil" Weyerhaeuser Jr., reported to the executive committee of the board of directors on postwar capital requirements. He said, "We should get as clear a picture as possible of how we can discharge our postwar job responsibility to the 1600 employees now in the military services, to the communities in which we operate and to those communities in which we own large blocks of timber which are now undeveloped and withheld from development by others." The estimate for capital required for the Coos Bay operations for 1945 to 1949 totaled $7 million. The total for the entire company was $47,747,000.

In 1945, the company opened an office over a drugstore in downtown North Bend. It was the first official company presence in the area. Working out of the three-room office were Conklin, Howard Henderson as land agent, Harold Taylor, office manager, and Mrs. Duffy, secretary. They stayed in this office until the mill was built, and the new office on the mill site was completed.

Bob Conklin, ever sensitive to the community's acceptance of Weyerhaeuser, wrote to Ingram in 1945 outlining his thoughts on local policies. His main points were the Coos Bay operations should be geared to the local communities; the mill could be sustained on a double shift capacity of 100,000 Mft. per year, and a management plan should be prepared to insure a continuous timber supply for such a plant. He urged that they should employ local people wherever practicable, and that returning service men should have first preference. He also stated the company should purchase locally, stocks and prices permitting. He also said the company should back sound community growth and progress.

Also by 1945, a logging plan had been proposed that finally settled the question of where to start, Millicoma-Lake Creek or South Coos. The decision was to cut for the first twenty-five years in the Millicoma-Lake Creek area. The logging would be clearcutting in compartments of one to three settings in size, the maximum area not to exceed 120 acres. An estimated 60 percent of the area would be logged on the first cutting with 40 percent being reserved as fire breaks and seed blocks. The interval between cuts will be gauged by the success of natural regeneration and in any event five years must elapse after the first cutting. Under this plan 86,400M board feet would be produced each year. With a 25 percent mill over-run this would produce 102,291M board feet of lumber.

Conklin recommended that the first mainline road be built under contract. He also stressed that the logging plans be flexible enough to take advantage of weather changes as they occur. He cautioned that, "Dirt roads must be made use of in dry weather with timber down and rigging ahead for quick moves to rocked roads when the seasonal rains occur. This entails some extra spur roads built ahead for spring and fall logging and timber down thereon to permit moves on short notice." For the next forty years the most successful loggers on the Millicoma were those who could stay on the dirt the longest.

Deep in the still unroaded forest something was occurring that was to determine all the future road building for many years ahead, but it was not yet evident to the foresters and engineers. The winter storms that lash the Oregon coast are almost an annual event, usually beginning in the first week of December. In the winter of 1949, and again in 1951, severe storms came howling off the Pacific and whipped through the Millicoma. No large swathes of windthrown trees resulted from the storm, but rather individual trees and small clumps went down throughout the forest. Lying in the shade with some of their roots still alive, these windfallen trees proved ideal breeding grounds for the Douglas-fir bark beetle, *Dendroctunus psuedotsugae.* Even as the first surveyors were laying out the mainline, flaming red patches of dead and dying trees were pockmarking the forest. The crews of the 1946 growth study had noted the mortality caused by the beetles, but something bigger was building up.

For the next several years, Conklin and his small crew in the office above the North Bend drugstore were incredibly busy. In 1946, he hired Royce Cornelius as branch forester and Carl Raynor as engineer for the mainline road. Conklin was involved in piecing together the property needed for the mill, securing booming rights on the river and keeping clear of the fights between rival lumbermen, Jim Lyons and George Vaughan on the South Fork. The biggest challenge was pushing the mainline road up the Millicoma. The road contract had been given to Morrison-Knutson, which was plagued with old equipment because of the war shortages.

Conklin was rewarded in the spring of 1949 when the company sent him to the executive management program at Harvard University's School of Business from February to June. He and his family had become well liked and respected citizens of the Coos Bay community. In October 1949, Conklin was moved to the company's Springfield, Oregon, operations, thus ending the Bob Conklin era on the Millicoma. In 1951, he left the company apparently after disagreements with his new boss at Springfield. Charles W. Fox, president of the Cascade Plywood Company had written to Phil Weyerhaeuser, the president of Weyerhaeuser, asking for his opinion of Conklin. Weyerhaeuser replied, "Bob Conklin has a fine personality. He is honest and has a driving ambition. I am disappointed if he finds nothing in our organization to satisfy him, since he represented us ably in the formation of our Coos Bay operation—especially in winning the friendship of the community. While at Springfield he was gaining experience in new lines, which came on top of Harvard Business School executive training course. Question mark seems to be satisfying his immediate superior. All in all in addition to liking him, I have respect for his capabilities."

Royce O. Cornelius arrived on the Millicoma in July 1946, after his service with an artillery unit in Europe. He had been the branch forester at Vail-McDonald in Washington where he had established some of most successful Douglas-fir plantations in the company. He was a tenacious firefighter and got along well with the loggers. A graduate of the University of Washington's forestry school, Royce was built like a rock, with powerful legs and torso. He had a resonant voice, a winning smile, and he loved practical jokes. He was a take charge guy. When he entered a meeting he was soon the dominant person in the room. Like Bob Conklin, Royce Cornelius must have thought he had a dream job when he was sent to Coos Bay in 1946. The forest lay unbroken waiting to be managed and the forester was there first. It was his to protect, regenerate, and harvest.

Getting logs out of the Millicoma had never been easy. Dow Beckham, in his book *Swift Flows the River,* chronicles the early logging on the Millicoma from Allegany to Matson Creek. Beginning in the 1870s, loggers used bull teams to put logs into the river where winter "freshets" would sluice them down. Beginning in 1884, a series of splash dams were constructed on the East Fork of the Millicoma. By the turn of the century, logging railroads were being built up some of the tributaries. In 1890, Henry Laird had built a railroad up what is now known as Woodruff Creek at the site of what became Weyerhaeuser's logging terminal at Allegany. The first truck logging on the Millicoma began at Brady and Neal's camp at Glenn Creek junction with the East Fork in the mid-1930s.

The geology of the southern coast range presented some formidable challenges to road builders. A million years ago the sea covered all of the land now covered by the Millicoma Forest. Sometime during the Pleistocene era a tremendous uplifting occurred which produced steep-walled canyons and streams began cutting their way through the massive deposits of sandstone. This was also a time of heavy precipitation, and as the steep slopes became saturated the land moved sometimes in large catastrophic landslides resulting in lakes such as Loon Lake. Landslides are still not uncommon today. Geologists have named the rocks of this portion of the coast range the "Tyee formation." The Tyee formation, as described in Baldwin's *Geology of Oregon,* is made up of a "rhythmically bedded, micaceous sandstone grading upward into siltstone." Geology had presented the road builders with not only steep canyons subject to landslides, but also with a soft rock that had little durability for road surfacing. The chief geologist of Oregon, John Eliot Allen, wrote to Conklin in 1944, "Our field work has shown fairly conclusively that the chances of occurrence of any hard rock suitable for road surfacing in the area in which you are interested are very small or not existent. The central portion of this part of the coast range consists of a great basin or downwarp made up largely of shales and sandstone through which the older, igneous hard rocks do not penetrate." It was little wonder that the road up the Millicoma and to the falls on Matson Creek became one of the costliest roads ever built by the Weyerhaeuser Company.

References

Baldwin, Ewart M. *Geology of Oregon.* Ann Arbor, Mich.: Edwards Brothers, 1959.

Beckham, Dow. *Swift Flows the River.* Coos Bay, Ore.: Arago Books, 1990.

Cornelius, Royce O. Personal communication. January 5, 1995.

Conklin, Robert P. Weyerhaeuser Company Archives, Tacoma, Washington. Record Group 3. Boxes 33, 39, 48, 52, 54, 66.

Oral interviews, 1976. Rex Allison, Del Hilliard. Weyerhaeuser Archives, Tacoma.

"14 miles of sheer rock cliffs and steep hillsides"

—Coos Bay Times reporter

4 / The Early Development

The mainline road along the East Fork of the Millicoma and up into Matson Creek was to be the artery over which millions of feet of logs would flow to tidewater. The tonnage that came down that road over the ensuing decades was incalculable. The builders knew that it would take a pounding. Ballast, drainage, and surfacing were critical.

Bob Conklin's budget included about $800,000 for the mainline road, which one *Coos Bay Times* reporter described as "14 miles of sheer rock cliffs and steep hillsides." After eight bridges, tons of blasting powder, and a mountain of rock, the final cost of the road had soared to between one and two million dollars. Some sections of the road had run as high as $160,000 per mile. Many people thought that the company had been taken advantage of by Morrison-Knudsen, which built the road on a cost plus basis. For years the Coos Bay operations were the largest purchasers of blasting powder and culverts in the entire company. The Millicoma Road was completed by the end of 1949.

Along with the road construction, the Allegany terminal was being built. The plan called for logs from the Millicoma to be dumped into the river at Allegany, rafted and towed by tugboat to the mill at North Bend or to the booms along the lower river and upper bay. Allegany was also to be the woods

(clockwise from top left)

Kelly Lookout construction: carpenters Bob Gould and Norris Baxter; tower raised by Ovie Coleman.

Bringing down a big one. Courtesy Weyerhaeuser Archives.

Mack logging truck coming down logging road on the Millicoma. Tons of blasting powder were used to construct these roads. Courtesy Weyerhaeuser Archives.

Loading out right-of-way logs. Courtesy Weyerhaeuser Archives.

A load going in, Allegany Terminal. Courtesy Weyerhaeuser Archives.

Boom boat making up a raft. Courtesy Weyerhaeuser Archives.

headquarters. By 1952, twelve family houses and eight small "apartments" had been completed, along with the large shop building, the woods office building, the forestry building with its hose tower, and a small tree seedling nursery. The unloading facility was a huge A-frame. A fleet of boom boats, known as "log broncs," would round up the logs after they splashed into the river. For many years a tug boat towing rafts of freshly cut logs was a familiar sight as they churned down the river past the dairy farms and fields along the North Fork of the Coos.

The village of Allegany across the river from the terminal consisted of the Allegany School and the Allegany store, which also contained the post office and the pay telephone. The Ott family, who ran the store and served as postmasters, was probably the most prominent family in Allegany. Their roots went back to the very beginnings of this tidewater outpost. The newcomers across the river came over to get their mail and to use the telephone. The terminal had no telephones to the outside world. Communications to the North Bend office were by radio.

One of the most influential newcomers to Allegany was Royce Cornelius, who not only built the Allegany terminal, miles of forestry roads, lookout

On their way to the mill. Courtesy Weyerhaeuser Archives.

towers, a radio relay station, and established the first plantations in the old Vaughan logging area, but he also got elected to the local school board. The old schoolhouse was about ready to fall down, and an Allegany resident, Charlie Jones, was determined that his six children would have a better school. Jones persuaded Cornelius to run for the school board, ran the campaign, and got him elected. In typical Cornelius style, a brand new school soon rose on the hill across the river.

Cornelius also began putting together his forestry staff. The first were Dean Higinbotham and Herm Sommer, who made major contributions to the management of the Millicoma Forest. Sommer was hired by Bob Conklin in 1948. Another hire was Norris Baxter, who was probably the finest rough carpenter in the country. Baxter could build anything from a toolbox to an eighty-foot lookout tower with incredible speed and consumate skill. Building the remote Kelly Lookout tested the skills of everyone connected with it.

One of the first jobs for Royce and his two foresters was opening up fire access roads. Most of these followed the ridges and had been pack trails for the fire patrol. They put in many a long hot day with Clyde Winslow operating an old D-7 cat. Since they mostly followed the contour lines, the road had many

Royce Cornelius with Bobbi Rhodes, the first lookout at Kelly Lookout. Courtesy Weyerhaeuser Archives.

switch backs and steep pitches. It was over this road that the long cedar poles for the construction of Kelly Lookout had to be hauled from an area near Vaughan Lookout. The site for this new lookout was Kelly Butte on High Ridge at an elevation of 2,500 feet, and some fifteen miles from Vaughan.

Ever resourceful, Royce borrowed a pole trailer from the power company. With the brand new Dodge Power Wagon in front and Royce steering the pole trailer in the rear, the caravan headed for Kelly. At the first sharp switchback the Power Wagon would be on one side of the curve, the pole trailer on the other, and the poles bridging the canyon. Not only were the switchbacks an obstacle, but the steep pitches required more power than the Dodge, and a new Ford flatbed truck that had been added to the parade, could muster. At one point a very frustrated Cornelius removed the governor from the Power Wagon. Dean Higinbotham remembers Royce flinging the governor off into the woods with "some final words from Royce." The team gave up on the pole trailer and had the shop make a pan for the top end of the poles. Clyde Winslow in his old D-7 then "walked" the poles out to Kelly Butte with only a broken track rail and a few thrown pads.

Once on the site, the next task was the raising of the four cedar poles that would support the lookout. Ovie Coleman, who learned his engineering in the woods, sized it up and with only the winch on the Power Wagon and block and tackle raised the poles. The structure soon took shape with Norris Baxter, Harold House, and young Tommy Graham clambering around like monkeys.

The infrastructure was taking shape. The forest awaited.

References

Conklin, Robert P. Weyerhaeuser Company Archives, Tacoma, Washington. Record Group 3. Boxes 33, 39, 48, 52, 54, 66.

Cornelius, R. O. Personal communication, January 5, 1995.

Higinbotham, D. D. Personal communication, 1996.

Weyerhaeuser, J. P., Jr. Weyerhaeuser Company Archives, Tacoma, Washington. Record Group 3. Box 53.

Youst, Lionel. *Above the Falls: An Oral and Folk History of Upper Glenn Creek, Coos County Oregon.* Coos Bay, Ore.: South Coast Printing, 1992.

TIMBER IS A CROP. DON'T BURN US UP

—Sign at Clemons Tree Farm 1941

5 / The Forest Gets A Name

In August 1949, Bob Conklin wrote to Walter DeLong, head of Weyerhaeuser's public relations department in Tacoma about the plans for dedicating the Millicoma Forest as a certified tree farm. Conklin wrote, "I am personally very happy that Mr. Weyerhaeuser has decided to hold no celebration or fanfare at this time. It is believed that the public reaction will be much more in our favor, if the tree farm dedication ceremonies occur shortly after construction has started on the sawmill property. We are particularly thin-skinned when the company is accused of using the tree farm movement as a publicity front. You and I know that our organization is conscientious in these matters and I am particularly desirous of demonstrating our sincerity here at Coos Bay."

Royce Cornelius never really liked the name "tree farm." He had been raised in a Forest Service town in Colorado, and his wife was the daughter of the Forest Supervisor. The Millicoma was a forest to Royce, not a tree farm. But the term "tree farm" had special significance to the Weyerhaeuser Company. In the 1930s and up to World War II, Gifford Pinchot, the fiery first chief of the Forest Service and a leader in the conservation movement, having despaired of ever getting management on industry lands, called for federal regulation of cutting practices on private lands. The Weyerhaeusers had been listening to the eminent forestry consultant David Mason, who was preaching sustained yield to industry clients, as well as their own foresters, who were telling them that the cut-over lands could grow trees if fire was kept out. For many decades there was little or no conscious reforestation on private lands. No prudent landowner or investor would plant trees if they burned up in a few years. As the largest private owner of forestland, the Weyerhaeuser Company was probably more concerned than other owners on what to do with its cutover lands. For awhile there was a Logged-Off-Land Department in Tacoma that attempted to interest midwestern farmers in buying these lands. It soon became apparent that even the toughest Minnesota farmer would have trouble farming these "stump ranches." During the Depression, thousands of acres went back to the counties for delinquent taxes. The drum beat for legislation from Washington was growing louder, which prompted George Jewett, a grandson of Frederick Weyerhaeuser, to say, "I feel strongly that unless the industry has a definite workable plan or course of action, we are going to be told what to do whether we like it or not."

The company eventually abandoned the Logged-Off-Land Department and formed the Reforestation and Land Department under C. D. "Dave" Weyerhaeuser. Company foresters, Ed Heacox, Bill Price, and Mike Grogan convinced management that they could develop a fire protection system that could hold losses to an acceptable level, and that tree growing would pay. A 130,000-acre tract in Grays Harbor County, the Clemons operation, was turned over to the foresters to practice what they had been preaching. The land, some of the most productive forestland in the temperate world, had been burned and re-burned for decades and was, for the most part, a barren brushfield. But *tree farm?* The name, probably the idea of Weyerhaeuser's ebullient public relations manager, Roderick Olzendam, was quickly adopted by the Weyerhaeusers as something the public would readily understand. Foresters Pinchot, Fernow, and Greeley had all talked about timber as a crop that could be renewed just like any other crop. So tree farm it was, and in ceremonies in June 1941 in Montesano, Washington, the Clemons Tree Farm was dedicated. The movement spread across the country through state and industry associations, and soon there were certified tree farms in almost every state.

The company was granted Tree Farm Certificate No. 100 on November 23, 1949, for the Millicoma, but in view of Conklin's concerns the celebration was not held until July 9, 1951. It was originally certified as the Millicoma Forest Tree Farm, but it is now known as the Millicoma Tree Farm. The event attracted statewide publicity with stories in the Portland *Oregonian* and the *Oregon Journal*. The *Coos Bay Times* ran features on July 2, 5, and 9.

(left) Carl Alvin Schenck dedicating the Millicoma Forest, 1951. Art Karlen, first manager of the Coos Bay operation, is seated behind him. Behind Karlen is William Greeley, head of the West Coast Lumbermen's Association and former chief of the U.S. Forest Service. Courtesy Weyerhaeuser Archives.

(right) Logging Superintendent Bill Gaines at dedication. Schenck and Karlen, seated. Courtesy Weyerhaeuser Archives.

The star attraction at the dedication was Carl Alvin Schenck from Lindenfels, Germany. Even at eighty-eight years old, Schenck was an imposing figure. One of the earliest foresters to practice in the United States, he was hired by George Vanderbilt to manage his estate at Biltmore in North Carolina. There he founded the Biltmore Forest School and was its director from 1898 to 1913. He then returned to Germany and served in the First World War in which he was wounded. On this trip, his last visit to the United States, he was feted across the country by his former students, "the Biltmore boys." However, this visit in 1952 was not his first to Coos Bay.

Beginning in 1910 Schenck had brought his students to Coos Bay each summer to see, first hand, logging and milling in the big timber. The C. A. Smith Lumber and Manufacturing Company, the biggest operator in Coos County in the early 1900s, had hired John Lafon Jr., one of Schenck's students, to build a nursery and try reforestation.

As part of the dedication ceremonies, Royce had laid out a sixty-five-mile field trip for the dignitaries, which included a trek out to Kelly Lookout over the dusty, winding access road. The trip must have terrified some of the townspeople who had never seen what lay in their "backyard." Schenck captivated

Bill Gaines, left, holding map of the Millicoma. Royce Cornelius at the microphone. Courtesy Weyerhaeuser Archives.

everyone with his vigor and enthusiasm and climbed to the top of the seventy-foot tower unaided. Among the speakers at Kelly Lookout was logging superintendent Bill Gaines, clearly uncomfortable in his role as a public speaker. Gaines told the assembled throng how the logs were cut, hauled to the landing, and then trucked to the dump at Allegany where they were rafted and sent to the mill. He amused the group when he concluded by muttering, "where, I understand they are made into lumber."

At McKeever Lookout, just above Allegany, the Carl Alvin Schenck unit of the Millicoma Tree Farm was dedicated with an emotional speech by the old forester. At one point, looking up at the sky, he said, "Look at the eagle flying over my tree farm!" No one told him that his eagle was a buzzard.

In spite of association executive Bill Hagenstein's car running out of gas on the way to Kelly, the event went off without any mishaps. It was well past eight o'clock that evening when the guests got back to Coos Bay. Even the locals had seen country that they knew little or nothing about. They had seen the forest that would become the dominant economic force in their community for decades ahead. The forest now had a name.

References

Conklin, Robert P. Letter to Walter DeLong. August 30, 1949. Weyerhaeuser Company Archives. Tacoma, Washington. Box 53.

Coos Bay [Ore.] *Times.* July 2 , 5, 9, 1951.

Hagenstein, William. Letter to J. P. Weyerhaeuser, Jr. November 25, 1949. Weyerhaeuser Company Archives. Tacoma, Washington. Record Group 3. Box 53.

Millicoma Tree Farm. Weyerhaeuser Company Archives. Record Group 7. Public Affairs. Box 4.

The Oregon Journal. July 10, 1951.

The Oregonian. July 10, 1951.

"This was a unique breed of men." —Bob Conklin

6 / The Logging Begins

In July 1950, the yellow crew buses and foremen's yellow pickup trucks snaked their way up the Millicoma road to the first setting just above the six-foot String of Pearls Falls on Matson Creek. It was here that Ovie Coleman, then the climber, topped and rigged the first spar tree. Herm Sommer, with Jim Bedingfield as his compassman, had type mapped much of the country where the first logging began. On July 16, 1950, the first load went down the road. A short distance above the falls, Conklin Creek joined Matson Creek. Over half a century ago, Dan Mattson had taken up a homestead just about where the first trees were felled by the Weyerhaeuser cutting crew. Mattson had sold out his claim and moved to Catching Slough on the upper bay in 1903. His only mark left on the land was his name on the map, although it was misspelled.

The first loggers were the fallers, and they must have rubbed their hands in glee to see this timber. At that time the cutters were paid by the scaled volume of timber they cut, not by the hour like all the other members of the woodworker's union. They were known as bushelers. The rate had never really been adjusted from the days of the cross-cut saw and ax. With the advent of the power saw, which began appearing in the woods just before the war, a pair of bushelers could increase their daily production enormously. In timber like

Ovie Coleman, high climber, topping a spar tree. Courtesy Weyerhaeuser Archives.

this, a busheler could make $50,000 a year, while a yarder operator was lucky to make a fraction of this. This inequity finally ended a few years later, and the cutters were put on an hourly basis.

Johnny Hewitt, Weyerhaeuser's first bull buck, was a rangy, sandy-haired, friendly man, respected by his crews. He didn't look like a logger; he wasn't especially big, wore glasses, and was a diabetic. But he had years of experience falling timber and running cutting crews. Hewitt could lay out strips for his crew so that they would not fall timber into another crew, and he preached safety constantly. Once the timber was down, a landing was laid out, and a spar tree selected. Ovie Coleman was the high climber and a showman. After climbing perhaps eighty to one hundred feet up the trunk of the tree, he would cut off the top and hang on while the tree snapped back and forth. Many tours in the early days for dignitaries from town featured a tree topping by Ovie. On at least one occasion after the top had been severed, Ovie, to the horror of the safety man, stood up on the top and waved his hat.

The first logging superintendent was Bill Gaines, an Irishman who liked his whiskey and was the quintessential logger. No mill he ever supplied ran out of logs, no canyon was too steep, no timber too big for Gaines. He knew rigging. With enough blocks, line, and tail holds he could probably move the world. He also knew loggers. Gaines, however, was not a planner or manager, despite the fact that earlier in his career he had owned a grocery store and sold cars. His desk at the woods office had little on it except dust, which he would sweep off with his big hands. He disliked making decisions, and one of his most often used expressions was "either way is all right." When under pressure from downtown or Tacoma, he would go off on binges and frequently disappeared for several days. He was probably the loneliest man in the Allegany camp; he was not accepted socially by the managers downtown, and the people in camp all reported to him, which made fraternization difficult.

Bill Gaines and his wife, Gladys; her son, Keith; and their dog, Queenie, lived in House #1 at the Allegany terminal. My wife, Irene, and I were their neighbors in House #2. Bill was a loving husband and a kind father to Gladys's son. When in his cups Bill could recite every verse Robert Service ever wrote. We can still remember Bill at our dinette table regaling us with "Dangerous Dan McGrew" while spewing cigarette ashes all over Irene's tablecloth.

Those first logs came down from the Millicoma on the brand new Mack trucks. The timber averaged 95,000 board feet per acre in those first settings, and the cut for that first year was 16 million feet. The mill was still not completed, so the logs probably went to the mill on the bay that was cutting the lumber used in constructing Weyerhaeuser's sawmill on North Bend's waterfront.

The large new mill going up on Coos Bay was big news for the community. Souvenir editions of the *North Bend News* and the *Coos Bay Times* told about the company's plans. The *Times* story pointed out that, "It will take about 100 years of growth before the area can be harvested again." On May 1, 1952, Ben Chandler, a pioneer Coos Bay area lumberman, pulled the whistle that signaled the start of the mill. The mill whistle had come from a derelict steamer wrecked on the bar years ago. As with any startup, there were some glitches. In a letter to the directors, Phil Weyerhaeuser said, "None of us went down to Coos Bay to see the startup pains. After the eighteenth log got up the hoist something busted. Guess it is going again now." Despite these glitches the mill did manage to put out 40,000 feet that first day. As the first log went into the mill, no one was thinking of the day in March 1989, when the last log would go through the mill, but in those thirty-eight years it was a going concern for sure. Lumber from the docks went to markets throughout the world. Coos Bay soon captured the title the "World's Largest Lumber Port." The first shipment was a million and a quarter feet of Douglas-fir loaded on the *George S. Long*.

Art Karlen, first manager of Coos Bay operations. Courtesy Weyerhaeuser Archives.

Bob Conklin left Coos Bay in 1949 and was succeeded as branch manager by Art Karlen. Where Conklin was a woodsman, Karlen was a millman. Earlier in his career he had been a dairyman and cheese maker near Grays Harbor in Washington. More important, he had served in France in the First World War with Charley Ingram. Karlen was selected by Ingram and was sent to Coos Bay from Longview, Washington. Karlen was bitterly disappointed that he didn't get the job to succeed the legendary Harry E. Morgan as manager of the company's huge Longview operations. It was a while before he finally built a house and moved his family to Coos Bay.

Karlen was a tough manager, but he was fiercely protective of his people, his mill, and his tree farm. He could go through the mill and know most of the people by their first names. He took the union almost as a personal affront, and Coos Bay was a tough union town. Karlen seemed to thrive on controversy, much of which he started himself. The Art Karlen era at Coos Bay saw some tumultuous battles with the Industrial Woodworkers Association (IWA), many of which led to strikes. When he retired in 1960, the *North Bend News* noted, "He is an outspoken man, a trait which he feels has sometimes got him into trouble . . . people basically like honesty and the direct approach. A union leader the other day said, 'We have had plenty of differences with Karlen and I think he is too hard-headed. But there is this much about Karlen, he says what he means and we know where we stood when he said something.' " Alma

Aerial view of the Weyerhaeuser plant in 1960. Courtesy Weyerhaeuser Archives.

Starvich, one of Karlen's former secretaries, said, "He was a delightful man, very frank, terrific sense of humor and just a wonderful person. He had a gruff exterior, but underneath was just the opposite."

George Weyerhaeuser was Art Karlen's counterpart for a while at Weyerhaeuser's Springfield, Oregon, operations. In the late 1950s, Weyerhaeuser and his timber people came over to the Pillsbury area of the Millicoma to look over the possibility of sending some of the timber east to Springfield's mills rather than west to Coos Bay. When Karlen got wind of this, he called Tacoma and said, "Look, I don't care what his name is, he's not going to get my timber!" And he didn't.

In the pre-dawn darkness, the "crummies," as the crew buses were universally called, wended their way through the silent streets of Coos Bay. Waiting at the corners were little knots of men dressed in Frisco jeans, stagged off just below the knees, and held up with suspenders. They probably wore "Romeo" slippers on their feet and carried their caulked boots slung over their shoulders. They wore a hard hat on their head, and if it was raining, which it frequently was for most of the year, they had rain clothes. Most important was the lunch bucket, packed by some sleepy wife, and jammed with at least three hefty sandwiches, pie, cookies, fruit and a big thermos filled with strong, hot coffee. This was a unique breed of men. Like miners, oil drillers, cowboys, and farmers they were wresting the wealth from the natural resources of the richest country in the world. These were not the Paul Bunyans of the old time logging camps. For the most part they were sober, family men. They lived at home, not in camps. And every day, as they went up into the far reaches of the Millicoma Forest, they were pitting their skills against the formidable challenges of one of the most dangerous occupations in America.

References

Coos Bay Times. May 1, 7, 1951; January 21, 1952.

Holbrook, Stewart. "He Can't Stay Out of the Woods." Advertorial for the Bethlehem Pacific Coast Steel Company in *The Timberman* (July 1951).

Morgan, Harry E., Jr. Taped interview with author, July 2, 1986.

North Bend [Ore.] *News.* January 24, 1952.

Weyerhaeuser, J. P., Jr. Letter to Directors, May 7, 1951. Weyerhaeuser Archives. Tacoma, Washington. Record Group 3, Box 53.

Youst, Lionel. *Above the Falls: An Oral and Folk History of Upper Glenn Creek, Coos County Oregon.* Coos Bay, Ore.: South Coast Printing, 1992.

"I will try to get every log every day that I can possibly get." —Ovie Coleman

7 / The People

By 1952, Weyerhaeuser had 272 woods employees, sixty-three of whom worked on the rigging. Another fifty were on the cutting crew. Truck drivers and equipment operators totaled seventy-three and construction had twenty-seven. Keeping everything running took another thirty-one at the shop and parts warehouse. Forestry and engineering employed twenty-two. Along with the mill employees, managerial and office staff, Weyerhaeuser was now the biggest payroll in the Coos Bay area and getting bigger almost every year. By 1962, the company employed 1,000 men and women, and the payroll had grown to well over $6 million. Along with the taxes paid to the county, state, and federal government, the Coos Bay operations had become a powerful economic engine. All of it depended on the Millicoma Forest.

Directing this crew of men were people like Art Blumenberg, Don Howell, Don Moore, and Ovie Coleman, the "side rods." Bill Hasbrook headed the construction crews, and Warren Browning ran the crushers. Everett Davenport was boss of the truck drivers. Homer Ednie ran the dump, and Lou Farris was the woods mechanic. Harry Spencer and Hank Reppetto were the engineers who laid out the miles of roads that were lacing the forest. In later years, Repetto referred to himself as the "roads scholar."

Woods Foremen at the 3000 line, 1955. Photo Arthur V. Smyth.

Downtown there were different people, a different society. They were the managers, the bean counters, the sales people, and, of course, the production men in the mill. Under Karlen was young Harry Morgan Jr., the son of the manager of Weyerhaeuser's largest operations at Longview and a rising star in the Weyerhaeuser hierarchy. Bill Jordan was the personnel manager and Karlen's right hand man in his wars with the union. Jim Opland was the comptroller, Bill Furher the sales manager, and Dudley Fosjord the purchasing manager. The first mill manager was John Snyder. Along side them stood some talented women, Peggy Champagne, Alma Starvich, Dorthea Yantis, and many others. Dorthea eventually succeeded Fosjord as purchasing manager.

Harry Morgan, at least on his first job at Coos Bay, was a part of the woods management rather than downtown. His first job was assistant to Bill Gaines, although Gaines really didn't want any assistants. He looked to John Wahl in Tacoma as his boss, not Art Karlen. This eventually led to Bill's downfall. Harry had the responsibilities for the cutting crew and the towing and boom crews. Harry admitted later that some parts of his job he knew very little about, but since it was a new operation everyone was learning together, and the relatively small group of people worked well together.

Rigging crew, February 1965. Photo by W. W. Dunlap.

Another of Harry's duties was safety, a subject that the company took very seriously. Hard hats were just coming into the woods, and getting the loggers to wear them was Harry's job. It wasn't easy. Before the hard hats, the loggers wore white or black cloth caps in the summer and heavy canvas waterproof hats in the winter. One of Harry's toughest converts was Ovie Coleman, then the company's only high climber. When Harry first ordered Ovie to wear a hard hat, Ovie said, "I can't wear that. I'll quit before I wear that." Harry said, "Goddamit, Ovie, you've got to. This is it. You're looked upon as a leader and if you don't wear it, nobody else is going to wear the damn thing, so you're going to have to." Ovie replied, "I can't. It gives me a headache." Harry persisted, "Try it. You won't get a headache." He came back the second day, "I've got a headache." Back and forth it went, but finally Ovie yielded. Years later, Ovie, in an oral interview, reminisced about the argument over hard hats. "Now I catch myself wearing it out in the yard mowing the lawn. It's been 26 years I've been wearing that tin hat. I have that first hat today that I had when I started at this operation."

Getting the boom crews to wear life vests was just as difficult. The first accidents were on the water rather than in the woods. Lou Hoelscher had become safety director and in the early 1950s went to Tacoma as safety director for the corporation.

The communities on Coos Bay were for many years isolated from any major metropolitan centers. Like all communities they had their social structure. The pioneer dairy farmers up the river had their club, "The Rivermen." The towns had their chambers of commerce and service clubs such as the Rotary, Lions, and Kiwanis. The major churches played an important role. The town's high school football and basketball teams had their loyal supporters, and during the 1950s Marshfield High's coach was a local hero as he turned out winning teams year after year. Steve Prefontaine, the brilliant and controversial runner, was probably the most famous athlete from Coos Bay. The school auditorium also played host to cultural events. The auditorium was packed for Leontyne Price. On another night, Arthur Feidler with the Boston Pops brought the cheering audience to its feet with his stirring "Stars and Stripes" finale.

For half a century, Weyerhaeuser, although a major landowner in the county, was little known in the Coos Bay area. Now, they had suddenly become the dominant economic presence in the community. Who were these "hot shots" coming down from Washington State? The local chambers of commerce and the press were elated at this tremendous boost in the economy of their towns. The North Bend newspaper, a weekly, was delighted that Weyerhaeuser's mill and office were in their town. The much larger Coos Bay paper, published by Sheldon Sackett, a liberal gadfly known throughout the state, was printed daily and faithfully reported the activities of the new giant on the waterfront.

Although most of the residents welcomed the newcomers with open arms, some were still a little suspicious of these big company executives. One evening some of the newly arrived executives were lifting a few at the Coos Bay Pirates bar, a local boosters club, when pushing and shoving developed between some of the Weyerhaeuser people and a prominent Coos Bay area businessman. At this point, the local, somewhat in his cups, told a Weyerhaeuser manager where he could stick a wire brush.

In 1952, Royce Cornelius was transferred to Tacoma headquarters, and Herm Sommer took over as Branch Forester. In the early spring of 1952, I arrived from Tacoma to take up my new duties as Inventory Forester. Driving up the Coos River on that beautiful day my bride of five months was at my side. She was to get her first look at her new logging camp home. At one time in her career, Irene had worked in the swank executive office of *Mademoiselle* magazine in New York City. When courting her I had told her, "Marry a forester and you will go far." Moving her from New York City to Allegany, Oregon, certainly fulfilled that pledge. On our honeymoon, driving down the coast from Tacoma to San Francisco in early December 1951, we had been buffeted by a severe winter storm with hurricane force winds roaring off the Pacific. Those same winds were wreaking havoc in the Millicoma Forest. The resulting windthrow added fuel to what was to become one of the most costly insect epidemics ever to hit the Douglas-fir forests of the coast range.

References

Coleman, Ovie, and Alma Starvich. Oral interviews with author, 1976. Weyerhaeuser Archives, Tacoma, Washington.

Morgan, Harry E., Jr. Taped interview with author, July 1986.

"God has an inordinate fondness for beetles."

—J. B. S. Haldane

8 / Chasing the Beetle

The British biologist, J. B. S. Haldane, credited the Creator for liking beetles because he made so many of them. In her book *Broadsides from the Other Order,* Sue Hubbell tells us that in the number of species, beetles represent the largest group of animals on the planet. She cites Terry Erwin of the Smithsonian, who estimates that there are probably ten million kinds of living beetles. Of this ten million there was one that changed all the logging plans for the Millicoma—the Douglas-fir bark beetle (*Dendroctunus psuedotsugae*).

In attacking Douglas-fir, adult beetles bore through the bark of the tree into the cambium layer, which lies between the bark and the wood. The first indication of a beetle attack on a tree are little piles of sawdust at the trees' base. Once into the cambium, beetles would construct fan-like egg galleries with a female and male in each gallery. The eggs would hatch into larvae, which in turn would pupate. The following spring a new crop of beetles would emerge from the tree, log, or windfall and attack surrounding trees. The tiny larvae or grubs would kill the tree by feeding on the cambial tissue. If the grubs were numerous enough the tree would be girdled, and a towering fir could be killed in a matter of weeks. The only defense the tree would have is its resin. If the tree could produce enough pitch it could flood the galleries and pitch out the

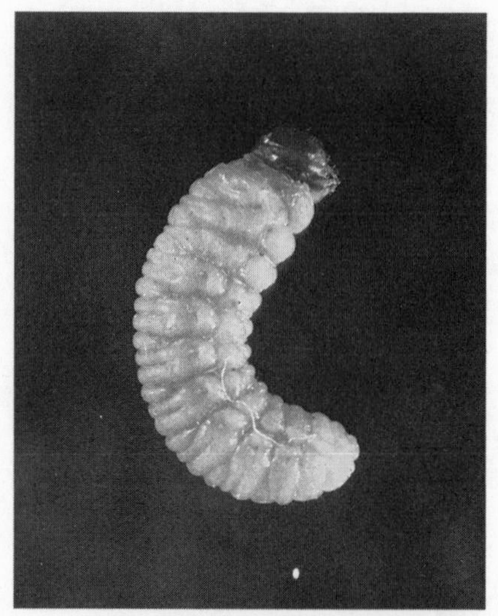

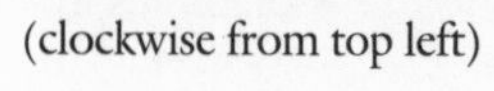

(clockwise from top left)

Douglas-fir beetle adult.

Douglas-fir beetle larva.

Beetle-killed tree. U.S. Forest Service photo.

Coastal Oregon blowdown, 1951 storm. U.S. Forest Service photo.

Blowdown: a breeding ground for bark beetles.

grubs. Many trees would have pitch streaming down their trunks, but few survived. As the tree died, the top would begin to fade from its deep green color to a pale yellow and then a flaming red. It was easy enough to spot the progress of the beetle through the forest. From the air it appeared as if the forest had a virulent case of the measles.

By 1952 the beetle-infested timber was showing up throughout the 180-year-old stands of the Millicoma. Between 1951 and 1954, the beetle killed an estimated 548 million board feet of timber. The task now was to salvage the dead trees before rot made them unmerchantable and to control the spread of the infestation. The key to both tasks was access. During the summer of 1952, a crash program of road construction was implemented in which crews were placed on a double shift basis. The elk and nocturnal denizens of the forest must have been startled when the darkness was lit up with the bright lights fitted on tractors bulldozing into their domain. By working before dawn and into the evening, sixteen hours of road construction were crowded into each working day. During 1951 and 1952, the crews pushed sixty miles of logging roads into the forest. The roads provided access into the heavily hit beetle areas so that the dead trees could be salvaged. In a three-year period, 62.5 million

Art Smyth, Branch Forester, examining a beetle-infested log, 1952.

feet of dead timber and windthrow had been recovered, but the beetle continued on its rampage well into the 1960s.

The road construction was, itself, creating problems. Some thirty million feet of right-of-way timber was being felled each year. This timber, which lay in shaded strips along each side of the right-of-way, provided ideal breeding grounds for a new crop of beetles. The felled and bucked timber in the clearcuts was not a problem since the logs would be exposed to the sun. It was found that timber felled from May to October generally dried out to such an extent by the following spring that it was not favored host material for the emerging beetles. These logs could be left in the woods for longer periods without the risk of adding to the infestation. Every effort was made to take out all blowdown and right-of-way logs by the following spring. Chasing the beetle had become a complex game.

During the beetle epidemic, the Millicoma Forest became a vast outdoor laboratory for an array of some of the top scientists in the region. A cabin deep in the forest at a locale called Mosquito Gulch became known to almost every entomologist and pathologist in the Pacific Northwest. This served as the headquarters for a team of government and company scientists attempting to find the answer to such questions as, why did the beetles attack one tree and not another? What attracted them? How far did they fly? What conditions did they like? How long would the dead timber last? What other fungi and insects follow the bark beetle? To track the spread of the infestation and locate the areas of heaviest mortality, miles of lines were run throughout the forest with plots established at regular intervals. Aerial reconnaissance flights were made over the forest, and for the first time, color aerial photography was used to spot the beetle kill.

In 1947, after the Coos Bay Growth Study indicated the mortality occurring throughout the 180-year-old stands, Bob Furniss, entomologist with the Pacific Northwest Forest and Range Experiment Station of the Forest Service, worked with Royce Cornelius on a cooperative beetle project. In the 1950s, the cabin and two wall tents at Mosquito Gulch were filled with people such as Norm Johnson, Paul Lauterbach, and Vaughan MacCowan from Weyerhaeuser's Forest Research Center and Ken Wright, Ernie Wright, and Toby Childs from the Pacific Northwest Station in Portland. These were not white-coated scientists working in a laboratory. Their days were spent with an ax, dissecting logs, felling trees, or busting through tangles of vine maple to establish study plots. In the evening at camp, low stake poker games went on under the hissing Coleman lanterns. Just getting into Mosquito Gulch, especially after a rain, was no mean feat. The road was a dirt access road that followed the old Fire Patrol trail, deeply rutted and slick as a greased pig after a rain. Mosquito Gulch was really misnamed; the mosquitoes were not that bad, and it wasn't a gulch. It was a bench at the very head of Fall Creek. It also had some of the finest timber on the Millicoma, which had attracted timber claims now lost to history.

The epidemic had forced the rapid roading of the forest, which certainly added to flexibility for its future management, but it was costly. Since road costs were amortized by the amount of timber that came out over them, the costs on the Millicoma were among the highest in the company because of the large amount of timber left behind in the race to reach the beetles.

In 1958, the Experiment Station published Research Paper 27, "A 10 Year Study of Mortality in a Douglas-fir Saw Timber Stand in Coos and Douglas Counties, Oregon." It was written by Ken Wright, the station's entomologist, and Paul Lauterbach of Weyerhaeuser's research department. The report pointed out that mortality during the ten-year period exceeded growth by more than 5 to 1. From 1951 to 1954, estimated mortality from wind and beetles totaled almost one billion board feet. In 1959, the *Journal of Forestry* published my paper, "The Douglas-fir Beetle Epidemic," which told the story of the salvage strategy. The beetle had now been immortalized in the professional literature.

References

Annual Forestry Reports: Millicoma Forest. Sub Group Forestry Box 5. Weyerhaeuser Archives, Tacoma, Washington.

Hubbell, Sue. *Broadsides from the Other Orders: A Book of Bugs.* New York: Random House, 1993.

Keen, F. P. *Bark Beetles in Forests: The Yearbook of Agriculture, Insects.* Washington, D.C.: Government Printing Office, 1952.

Smyth, Arthur V. "The Douglas-fir Bark Beetle Epidemic on the Millicoma Forest." *Journal of Forestry* (1959).

Wright, Kenneth, and Paul Lauterbach. "A 10 Year Study of Mortality in a Douglas-fir Saw Timber Stand in Coos and Douglas Counties, Oregon." Pacific Northwest Experiment Station. Research Paper 27, 1958.

"I wasn't sent down here to have people out on the bricks." —H. E. Hunt

9 / Changes in the Forest, Changes in the People

The unbroken expanse of forest that faced the first logging crews in 1950 had, by 1960, been changed radically by the human hand and the forces of nature. The main roads were built up the major drainages. Spurs were pushed up the draws and up onto the ridge tops. The pattern resembled a human circulatory system with arteries, veins, and capillaries. A four-digit numbering system had been devised so that not only could you find your way but also for accounting purposes. For example, the 3000 line ran up Matson Creek to the Lake Creek divide. The first spur off the 3000 line was the 3010 and a spur off the 3010 was labeled the 3010A.

Chasing the beetle had resulted in a patchwork of small clearcut areas surrounded by standing timber. Once the timber had been removed from these relatively small clearcuts, logging slash was burned, but not always. Herbaceous vegetation would spring up and the clearcut areas soon became prime feeding grounds for the rapidly expanding herds of Roosevelt elk. Game biologists commented that if they had set out to raise elk they could not have

Elk trapping on the South Fork. Photo Arthur V. Smyth.

Oregon Department of Game biologists tagging baby elk. Courtesy Weyerhaeuser Archives.

come up with a better plan than what Weyerhaeuser was doing. Soon tree damage from elk became a concern. Like the beetle epidemic, the elk problem attracted the attention of scientists. Soon state, federal, and company wildlife biologists were studying the problem.

Beginning in 1968, the Research Division of the Oregon Game Commission (later named the Oregon Department of Fish and Wildlife) studied the Millicoma Roosevelt elk. For the next nine years, game biologists investigated the elk's range, food habits, breeding habits, and damage to Douglas-fir seedlings. Funded by the Federal Aid Project, reports were prepared annually from 1968 to 1976 by biologists James A. Harper, William W. Hines, and James Lemos. Elk were shot with tranquilizer guns and tagged, belled, or outfitted with collars.

The most ambitious project was the construction of an eight-feet-high elk-proof fence, which enclosed 462 acres of elk range on the Millicoma. For the entire nine years of the study only one cow elk and her calf managed to find a way out of the enclosure. During the 1968 hunting season an unknown hunter removed a yearling bull elk.

By the end of the study, the most intimate details of the ecology of this magnificent game animal were now known. The ideal habitats for elk were clearcuts spaced no farther than six hundred feet from a cover edge. Trailing blackberry, grasses, and false dandelion were the leading forage plants. Douglas-fir seedlings that were browsed once released grew at a faster rate than those not browsed. They also concluded that big game animals caused only 5 percent of the study loss.

In 1955, the Millicoma was opened for elk and deer hunting. As opening day approached, hunting fever in the community intensified. Sporting goods stores were doing a land office business. The gate at Bridge 8 on the mainline road was to be opened at 5:00 A.M. on opening day. After the last log truck had come down the road on Friday night, hunters were allowed to drive up the road and line up behind the gate. Hunters slept little that night as they swapped stories and argued about the best places to go. At the gate each vehicle was tallied in, and the driver was issued a numbered card that he had to return by 6:00 P.M.

An army would pour through the gate and spread out through the now accessible forest. The first weekend some 190 bull elk had been tallied through the gate. Considering that these animals were huge, averaging 600 lbs., most had to be quartered and packed out of some unbelievably rough country. This was a remarkable spectacle indeed.

But the young trees managed to survive, and few if any logged areas were not fully stocked with young Douglas-fir seedlings. Although there was seemingly an abundant seed source for natural regeneration, relying on that was still a gamble, because a good cone crop only came around about once in seven years. The first planting following the 1950 logging took place in 1953 on the 1000 line up Conklin Creek and then on the 1040 line. In 1956, the company began aerial seeding with helicopters using treated seed to repel rodents. With planting crews and helicopters quickly following behind the yarders, a new forest was being established behind the harvesting of the old. By 1960, ten years after logging began, one billion board feet had been sent down the river. By 1970, the book inventory of the Millicoma Forest was still over 7 billion feet. Obviously the Coos Bay Growth Study had grossly underestimated the volume of timber on the Millicoma Forest.

There were changes among the people managing the forest as well. In 1955, Herm Sommer was appointed assistant logging superintendent. Downtown, Harry Morgan was appointed administrative assistant to Art Karlen. I succeeded Herm as Branch Forester. Herm Sommer and Bill Gaines were completely different people. A graduate of Oregon State and Yale, born in Chicago, Herm was meticulous, organized, and possessed with a Germanic thoroughness. He was careful with his money as well as the company's. He was strik-

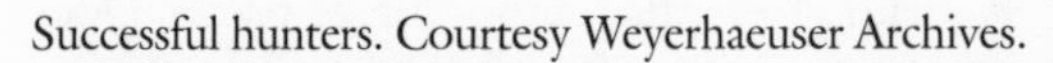

Successful hunters. Courtesy Weyerhaeuser Archives.

Harry Morgan. Courtesy Weyerhaeuser Archives.

ingly handsome, dark complected with a flashing smile. It was no wonder that this young, handsome bachelor living with his folks in Coos Bay was considered a catch. The wedding of Herm Sommer to the daughter of a prominent family was a major event on the social calendar of the Coos Bay area. Herm had a passion for building roads and later became a successful woods manager at the company's operations at Twin Harbors and Longview in Washington. Later in his career he was elected president of the Pacific Logging Congress. But working as an assistant to Bill Gaines in 1955 must not have been the easiest of jobs. That was about to change.

Art Karlen was becoming increasingly upset with Bill Gaines, who was still going through John Wahl in Tacoma rather than letting Art know what he was up to. This was no way to work with Karlen. It finally came to a boil in late 1957 when Art informed Wahl that he was going to fire Gaines. There were some heated words, but Karlen ruled Coos Bay. Harry Morgan had the unpleasant job of coming up to Allegany to tell Bill that he no longer had a job at Coos Bay. Some of us who knew what was going on watched Harry pull up in front of the woods office and go into Bill's office. Shortly after Harry left, Bill came out, jumped into his pickup and roared up to his house. He called John Wahl from town. John could no longer protect him, but he did get Bill the job at the company's Klamath Falls operations building a major reload station. The Gaines chapter had ended.

Clyde Kalahan. Courtesy Weyerhaeuser Archives.

Clyde "Sparky" Kalahan was yet another Longview "graduate" to be sent down to the Millicoma. He was the son of the longtime railroad dispatcher of the company's Longview operations, hence the nickname. A former Marine officer who saw combat in Korea, he still wore a short Marine haircut. He was a logging engineering graduate of Oregon State University, and Karlen hired him to succeed Bill Gaines.

His arrival at Allegany was viewed by some with trepidation. On his last job at Longview, where he was the assistant logging superintendent, he had the reputation of being a pretty tough manager. Shortly after his arrival, he fired the shop superintendent, ordered all the pickup trucks to be parked neatly with the front end out, and challenged the Forestry Department to prove they were not 8 to 5 pencil pushers. He replaced the shop superintendent with a retired Forest Service engineer, Harold Brischle. No one at Allegany called Kalahan "Sparky."

Beneath the tough exterior was another Clyde Kalahan. He had a delightful sense of humor, and after questioning what the foresters were doing, he became an active supporter of the forestry program for the Millicoma. He was an indefatigable worker, and although his desk may not have had much more on it than Bill Gaines's, it certainly was not covered with dust. He was a planner, and he left his mark on the Millicoma Forest.

That mark was the opening of the South Fork. In October 1944, Bob Conklin had written to Minot Davis, "You will agree, I am sure, on one basic conclusion, that is, that our logging area is divided into two natural units. Certainly it would be unsound to plan for simultaneous development of both of these divisions." In September 1945, Conklin wrote Al Raught in Tacoma, "Logging activity shall be conducted for the first 25 years in the Millicoma-Lake Creek areas." It was now a scant ten years since logging had begun on Matson Creek, but Kalahan had concluded that the time had come to open up the South Fork.

Kalahan had decided that it was time to stop chasing the beetle. He also felt that the log dump at Allegany was way over capacity and that dividing the operation, coupled with some changes in supervision, would help alleviate some of the problems with the union. In a Forestry Annual Report, I had written, "In order to convert the Millicoma Forest to a normal, managed forest we must cut in our mature stands for almost 60 years. Our mature timber stands will not stand still during those 60 years. They will either grow or deteriorate. It is our job as foresters to determine where each condition is occurring and how best to treat these stands for maximum production." We established priority logging classes. The high priority stands were mostly on the east side.

Just as Conklin had predicted, road construction up the South Fork was difficult. In early 1960, construction foreman Fred Thorne was laboriously chipping his way through the rock at Fall Creek headed for Williams River. The costs were soaring. One day Kalahan commented to Thorne, "Fred, can't you move a little faster? The alders are growing up behind you." An alternative for the South Fork timber could have been up the 5000 line and into Allegany. With the dump at Allegany already at capacity, Kalahan knew this was not the way to go. A new terminal would have to be built at Dellwood at the head of tidewater. The budget was approved in August 1960, and construction started soon thereafter. A year later, the budget for the shop was approved. Hank Reppetto tells how he and Harry Spencer argued over what elevation to use for the shop. Reppetto held out for a higher elevation than the original plans called for. He prevailed and was vindicated when the devastating Christmas week floods of 1964 inundated Dellwood but never reached the shop. The first load went into the new log dump on January 18, 1962.

Running the logging operations on the Dellwood side was Rex Allison, a forestry graduate of Washington State University. Rex came from the logging community of Lebam in the Chehalis River country of Washington, and he knew logging and loggers. Rex was production oriented, and his goal was to get more loads and more wood down the road to Dellwood than anyone at Allegany. Trying to reach Allison to inform him of woods shutdowns for low

Howard Hunt. Courtesy Weyerhaeuser Archives.

humidity or wind was always difficult. We suspected Rex turned off his radio so that he could get a few more loads. He had the respect of his men, and union troubles were few.

Some years later in an oral interview, Rex had this to say about the union:

> I think you have to give full credit to the union for the part they played. They have certainly been a factor in the development of the Coos Bay unit from an organization that was on strike 50 percent of the time it seemed like. I think there was more strikes than what there was anything else, but they have in their own way more or less forced a lot of better training, better equipment, better practices, until today they have evolved along with the company to the point where we lose very little time from work stoppages. We are able to negotiate the areas where we feel we need to be improved on. I think they are demanding on our supervision and have done as much to create better supervisors than any other single factor that might be involved.

Changes also took place downtown. In 1960, Art Karlen retired and was succeeded by Howard Hunt, an engineer whose last job had been at the plant in Everett, Washington. In 1958, Harry Morgan left to become branch manager at the company's Snoqualmie Falls operations in Washington. Hunt saw one of his chief jobs as repairing the tattered relations with the union. He said,

"I wasn't sent down here to have people out on the bricks." The union's leadership was courted by Hunt whereas Karlen ignored it. Nevertheless, in 1963, operations were shut down for sixty days because of a regional dispute.

It was during this period that Kalahan and I had a series of adventures. One occurred on an inspection trip to the Callahan area in Clyde's pickup truck. This entailed fording Cedar Creek, usually a fairly placid, mid-sized stream. However, this was early spring, and the creek was running higher than usual. But in we drove, and halfway across we stalled. We could not go back or forward—or anywhere. Also our radio would not reach anyone from down in this hole. After a couple of "oh shit!"s, Kalahan said, "Well, Artie, what do we do now?" Just then we spotted an unoccupied Jeep parked across the stream. Kalahan took off his pants, waded across, and found the keys in the ignition. Obviously a trout fisherman was somewhere downstream.

I was to stay with our pickup while Clyde borrowed the Jeep and went up to Cliff Woods's homestead some five or six miles up the Callahan road. Woods was a local logger and had lots of equipment with which to get us out. The fishing must have been good, because the Jeep owner did not return before Kalahan arrived with the Woods boys who got us out.

Then another incident occurred over in the Callahan area, again in Kalahan's pickup. Going down an old fire road through a stand of huge, old growth fir, we came to a windthrown fir some six foot through. Someone, who must have had a lot of time, had whittled a way through the windthrow with an ax, so in we went. We stopped, realizing that couldn't go forward or back. We were caught in a vise. Again, "Well, Artie, what do we do now?" We could not open the doors of the pickup. Kalahan was too big to even squeeze out of the window. Being a trim 120 lbs., it was obvious who had to go out the window. This I did, got the ax from the toolbox in the bed of the pickup, and with some very catty ax work chopped us out.

And yet another incident occurred. A fire in Brummit Creek to the south of our main holdings had started in Georgia-Pacific's lands and had burned into some of our timber. I was sent over to observe the work on the fire line, sketched the burn, and returned that night. The next morning I reported to Kalahan. Several days later, he said he wanted to see it, so off we drove. When we got to the fire, which was now out, we came to a cat road that had been pushed around the backside of the fire. Kalahan asked if the road went through and I said: "Well they were working on it when I was here, and I assume it does, but I don't know." Ahead we went, down and down. I think we both realized by then that if the road did not go through we would never be able to climb out of this hole. At the bottom of the grade there was a small trickle of a stream to cross. We did not make it. We were good and stuck. Again, "Well, Artie, what do we do now?" We left the pickup and proceeded across the creek to find, to our dismay, that that was as far as the "road" went. We climbed up the hand-built fire trail, and at the top came upon a Georgia-Pacific tanker crew with a radio. Kalahan asked if there was a cat nearby that he could borrow. Kalahan could drive a cat, but I could not; in fact, some people thought I couldn't drive a pickup. At any rate, they called their boss. His first question was, "Where did you say they were?" The next was, "What were they doing down there!" Kalahan was squirming. The logging boss for Georgia-Pacific was having fun with the head logger of Weyerhaeuser. His next comment was, "Tell him we are building a logging road in there next spring." We finally got together, and they understood our plight. They told us they would move one of their cats over there the next morning and get out Kalahan's pickup. Everett Davenport came over and drove us home. The next morning I was in Clyde's office waiting to go over to Brummit Creek when the phone rang. (We now had telephone service.) I could hear Clyde say, "It did? Oh no! Okay, Art, we will go over and get it." His face had turned white. Georgia-Pacific's lowboy carrying their cat had turned over on a curve, and the cat had plunged into a draw. No one was hurt, but we would have to get our pickup out ourselves. So we loaded one of our cats on to a lowboy, Davenport drove the truck, and we all went back to Brummit Creek.

References

Brown, Donna. *History of Southwest Oregon Region*. North Bend, Ore.: Weyerhaeuser Company, 1988.

Federal Aid Progress Reports. Millicoma Roosevelt Elk Research Project. Portland: Oregon Department of Fish and Wildlife, 1970–76.

Harper, James A. et al. *Ecology and Management of Roosevelt Elk in Oregon*. Portland: Oregon Department of Fish and Wildlife, 1987.

Kalahan, Clyde R. Personal correspondence with the author, 1995.

Weyerhaeuser Company Archives. Annual Forestry Reports. Record Group 7. Subgroup Forestry. Box No. 5. Tacoma, Washington.

"It is either a big flood, or a big snow, a big dry spell, a big fire. There is always something big about it and it seems to run in 4 or 5 year cycles." —Rex Allison

10 / Wind, Water, and Fire

Once again the forces of nature changed the logging plans on the Millicoma. It was Columbus Day, October 12, 1962. I was in my office at Allegany when about 11:00 A.M. I received word from downtown that a hurricane with winds in excess of 100 miles per hour was headed our way. It was estimated that the storm would make landfall at Coos Bay at 2:00 P.M. The rush was on to get everyone out of the woods. As usual it was difficult to locate Rex Allison, but soon everyone was headed in. Around noon the sky at Allegany had taken on a strange yellow cast, and there was an eerie silence. Not a breath was stirring. Just about 2:00 P.M. on the button, the tops of the big residual old growth firs, left from the turn-of-the-century logging on Packard Creek, began to explode. These trees were on a high ridge above Allegany, and up there the sky was soon filled with flying debris. The wind had not yet got down to Allegany, but soon we were hit. It became difficult to even stand up. In the woods the miles of cutting lines produced by the patchwork logging were especially vulnerable as were the pre-logged and sanitation salvage areas that I was so proud of. The trees were going down in swathes. Downtown the new plywood mill was under construction and the 4" x 8" plywood panels that sheathed the roof were being peeled off and sent spinning through the air like playing cards. Miraculously no one was injured.

But Hurricane Freida was just picking up steam. All night she roared northward, wreaking havoc throughout western Oregon and Washington. Freida smashed roofs and downed power lines in the cities and blew down some eleven billion feet in the forests of Oregon and Washington. Vicious winter storms are not at all uncommon to the Pacific Northwest coast, but no one had seen anything like this. With some five to six million feet of blowdown on the Millicoma Forest, there was the threat of another bark beetle epidemic. But prompt cleanup prevented a reoccurrence of a beetle build up. Now the loggers were chasing blowdown instead of beetles.

The year 1962 was a recession year, and lumber and plywood markets were in a slump—nothing new for such a cyclical industry. But what was new was eleven billion feet of windthrown timber that had to be salvaged as quickly as possible. Dumping this amount of volume into an already depressed market would put many mills out of business. Either the domestic market would have to recover, or new markets had to be found. The industry looked to Southeast Asia.

Sea otter pelts to China were the first exports to Southeast Asia from the Pacific Northwest. As early as 1788, traders included spars in the cargo and found a good price for them in China. John Meares built his ship *Northwest America* at Nootka in 1788 and wrote, "We also took on board a considerable quantity of fine spars for the Chinese market where they are very much wanted and of course proportionately dear." John McLoughlin's sawmill, which he built in 1827, began exporting lumber and spars to the Hawaiian Islands. Japan for years had been an export factor in Asia from its extensive domestic forest stands, but by the 1890s Japan's accessible forests had been pretty much cut out. In 1893, the American consul in Osaka, Japan, wrote in a report on the American lumber trade in Japan. It stated, "The trade in American lumber is increasing and bids fair to continue to do so, as it has been determined that the American production can be laid down in the open ports of the Empire more advantageously than the native product." Until Pearl Harbor "Jap squares" had been a common export to Japan. But in 1962, the U.S. began trading logs to Japan. Log yards soon sprouted up on the Coos Bay waterfront, and huge log ships became regular visitors to the port of Coos Bay. In 1964, the first logs were shipped to Japan from the Millicoma. In 1965, Weyerhaeuser loaded the *Kure Maru* with the first shipment of Douglas-fir chips. Export logs and chips proved to be a very lucrative market for forestland owners. The industry dependent on public timber, however, fought hard for restrictions on log exports. It became, and still is, a hot political issue. Exports from federal and state lands were eventually banned. But if the export trade in logs had its beginning in the wake of the Columbus Day hurricane, it was, indeed, an ill wind that bloweth good to some men.

Winter rains are a way of life on the Oregon coast. The rain gauge at Allegany frequently measured 80 inches or more a year. In 1953, the Millicoma was pelted with 117 inches. Local historians have recorded freshets and floods for years. In his history of Coos and Curry Counties, Orvil Dodge speaks of the great floods of 1861–62. In December 1879, yet another flood rivaled those of 1861. There were floods in 1885 and 1890. In 1953, floods inundated the Allegany log dump and sent houses down the West Fork and under the bridge at Allegany. In 1964, just before Christmas, a wild winter storm with torrential rains hit the Oregon coast. It wiped out the booms at Dellwood and sent logs hurtling down the river. It also destroyed three bridges disrupting logging plans again. Harbor Tug and Barge owned by Ray Beaudry and Bill Egenolf had the contract for towing logs to the booming grounds and to the mills. During the floods and freshets, Egenolf, who was a tug boat skipper, was busy with his boats trying to keep the escaping logs from going out to sea. Sometimes he didn't make it. Once Bill was overheard on the radio telling one of his boats to make, "a big Swedish swoop." During those wild days and nights there was a lot of Swedish swooping.

Although rain was plentiful in the winter, a lack of it in the summer was a way of life on the Millicoma. The threat was fire, not flood. Fire season could run from June to October, but usually the dangerous months were July and August, when little if any rain fell most years. The year 1952 was such a year, the dryest in a decade. A 1,700-acre fire erupted in the Williams River drainage. It crowned one night in three-hundred-year-old fir and also destroyed hundreds of acres of twenty-year-old Douglas-fir that had reseeded an old burn. The old growth was salvaged, but the twenty years of regrowth were lost. The year before, a much larger fire burned over 17,000 acres in the Hubbard Creek fire, below the Reston Rim. Some 3,000 acres of Weyerhaeuser land burned in that fire. The Douglas County Fire Protection Association from Roseburg headed the effort, and hundreds of local loggers worked on the fires. Royce Cornelius from the Millicoma took over much of the strategy, especially on the Williams River fire. This was my first big fire on the Millicoma, and I was responsible for the north line of the fire. With an army of tractors behind me, I laid out the fire line along a ridge. It soon grew into a wide road. To this day the road that started out as a fire line is shown on the maps as Smyth Boulevard.

Dean Higinbotham, a young Oregon State forestry school graduate, was the Millicoma's first protection forester. Dean was a bachelor, and during the early 1950s, he lived in the forestry building at Allegany. His domain was the miles of fire hose, the fleet of tanker trucks, the hundreds of fire tools, the radio system, and the lookout towers. Day or night Dean was always there. The state law required every logging side to have a water tank, a pump, and a toolbox. Weyerhaeuser's equipment far exceeded the state's requirements. Dean also had to keep an eye on the company's logging contractors, who generally had the bare minimum. The state law also required logging operations to shut down when humidity levels reached 30 percent or lower. So Dean was also the Millicoma's weatherman. He maintained the weather station at Allegany, and pyschrometers and fuel sticks were watched throughout the forest in periods of critical fire weather.

An event that occurred well over three decades ago will be long remembered by all who participated. It was late afternoon on a hot, dry summer day. The loggers had left the woods hours before. I was in my office at Allegany when suddenly a high pitched voice screamed over the radio, "Kelly Lookout. Help! Help!" I leapt out of my chair and called back to Kelly. There was no reply. I called Dean who was on patrol in his pickup. He came back immediately, "I heard it. I'm on my way." Repeated calls to Kelly from both of us raised no one. I then called McKeever Lookout who had contact with Kelly on the Fire Patrol's frequency. They called back shortly and told us that the voice we had heard was the lookout's thirteen-year-old brother who had been staying with his sister. The boy had called for help when his sister had passed out and fallen to the floor of the lookout tower. Dean finally reached Kelly and reported back that it looked like she had an acute case of appendicitis. Somehow he got the pain-stricken girl and her terrified brother down from the seventy-foot tower and into his pickup. He headed for Allegany. Meanwhile I had called for the ambulance from town and notified Keizer Hospital in North Bend that we would be bringing in a patient from Kelly Lookout with what appeared to be acute appendicitis. I met the ambulance at the first gate and escorted it up the mainline where we rendezvoused with Dean. We transferred the groaning young woman and her brother into the back of the ambulance. I went ahead of the ambulance leading him through the gate and down the road to town. At some point I noticed that the ambulance was no longer behind me, and I pulled off in a farmyard to wait for it. I was just about ready to turn around and go back to see what had happened when it finally came. The driver slowed and shouted out the window, "She just had a baby! Call the hospital!" My radio had gone dead. I ran up to the farmhouse and pounded on the door.

When the lady answered the door I said, "Please call the hospital and tell them that the appendicitis patient we were bringing in from Kelly Lookout just had a baby." She looked at me blankly, and I repeated my message. She called. At the hospital, mother and baby were fine, the young ambulance driver was ecstatic, and Dean Higinbotham was in shock. We subsequently learned that on the way to town the girl's brother starting screaming at the driver, "My sister's having a baby!" The driver pulled off the road, near panic. He had never been anywhere near a childbirth. He stopped a passing pickup truck, and serendipity triumphed. It was "Heavy" Johns, a local character who worked in the woods for Weyerhaeuser. "Heavy" knew all about childbirth, as he and his wife had about six kids. Together they delivered the baby. "Heavy" cut the cord with a pair of tin snips from his truck.

We were all nonplussed. How had a young pregnant girl been able to hide her condition from the Fire Patrol's chief warden who hired her for one of the most remote lookouts in the county? His only excuse was, "Well, she was a little plump." We surmised the family had been well aware of her condition, which was why the young brother had been sent to stay with her. Kelly Lookout had not seen such excitement since Schenck climbed to the top at the age of eighty-eight. For some period, Dean, the confirmed bachelor, was ribbed about his diagnostic skills.

Dean Higinbotham went on to other assignments and made the first timber sales on the Millicoma, and as raw material manager was responsible for export log sales. Floyd Page, Bob Wiggins, Ralph Sweet, and others succeeded Dean in the task of protecting the Millicoma Forest from fire. After the first years, no fire destroyed larger than fifty acres until the Ivers Peak fire in 1966, and that one flirted with disaster.

It was a warm, lazy August afternoon. The Allegany crews were on vacation, and the Dellwood sides operating in the Fall Creek drainage would start coming in shortly. Suddenly radios came to life with the alarm, "Fire!" The call had come from Bob Lehman, the loader operator, on the 6110. The fire was spotted by the hooktender, Chuck Allender, in felled and bucked and fresh slash adjacent to the area being yarded downhill to the landing. Lehman and his rigging crew ran to the fire, which was only a few feet across at this point. They pumped water from a pump can and used their hats and hands to throw dirt on the blaze. A line of hose was laid up the hill from the tanker on the landing, but the fire got away and roared up the steep slope, north to the 5040. Within half an hour from the sighting of the fire at 3:50 P.M., tankers were on the 5040, and hose line had been laid to attack the fire's rapidly advancing front. By 5:00 P.M. the entire 5040 was ablaze. The hose lines were abandoned,

Ivers Peak Fire, 1962. Courtesy Weyerhaeuser Archives.

Part of Ivers Peak burn. A new forest as far as the eye can see. Photo Robert Hanson.

and the tankers escaped to the east, all but one of which had to be left. On the 6110 where the fire had originated, crews with tankers and cats were attempting to confine it, but it was spotting everywhere. One fire was now half a mile to the west in the timber. That evening a tanker plane from the Medford airport dropped two thousand gallons of fire retardant on the flames with little effect. The crews made nine additional drops with planes and helicopters in the next two days with varying degrees of success.

At the height of the fire, over three hundred men worked the day shift, and one hundred worked night. Many of the supervisors worked day and night with little or no sleep. The crews came from everywhere. The company's Springfield, Oregon, branch sent a one-hundred-man team. The sawmill crew was called out as were crews from the local contract loggers. Men from the state's prison inmate crew were responsible for holding the east side. For a week it was touch and go, and mop-up and patrol went on for months. Some 1,760 acres of green timber, slash and felled and bucked went up in flames. What caused this blaze? Only one man in the rigging crew smoked, and he had been warned about flipping his cigarettes. He quit the day after the fire and left town.

Our forest, which began with fire, survived this one. Today, in the area where hundreds of men labored in the smoke and heat, a green blanket of Douglas-fir covers the landscape as far as the eye can see.

References

Brown, Donna. *History of the Southwest Oregon Region*. North Bend, Ore.: Weyerhaeuser Company, 1988.

Carey, Charles H. *General History of Oregon*. Portland, Ore.: Binford and Mort, 1971 [1922].

Cox, Thomas R. *Mills and Markets*. Seattle: University of Washington Press, 1974.

Dodge, Orvil. *Pioneer History of Coos and Curry County*. Salem, Ore.: Capital Printing Co., 1898.

Miller, Julian F. *Investigation of Ivers Peak Fire*. Coos Bay, Ore.: Coos Forest Protective Association, 1966.

Pettey, R. A. *Notes on Investigation of Ivers Peak Fire*. Coos Bay, Ore.: Coos Bay Branch, Weyerhaeuser Co., 1966.

"Forestry is exactly the same as agriculture. It is the application of superior knowledge and skill to produce a wood crop." —B. E. Fernow, 1895.

11 / High Yield Forestry

The 1961 Annual Forestry Report for the Millicoma Forest pointed out that with an eighty-year rotation, a cut of 165 million feet per year could be sustained. Under this plan, the present inventory of ten billion feet of mature timber would last sixty-three years. It became apparent because of the unbalance age distribution of the forest a continuous sustained yield would require holding the timber for an extremely long period with a relatively small annual cut.

By 1964, 30,576 acres had been depleted and the cut that had hovered around 165 million a year was now at almost 300 million, which included stumpage sales from the Callahan area of 58 million. It was obvious if the cutting continued at that rate the forest would not be on a sustained yield basis. The question was being raised, "should it be?"

When it came to questions of growth and yield, the "bible" of Pacific Northwest foresters had for years been the "Technical Bulletin No. 201, The Yield of Douglas Fir in the Pacific Northwest." It was written by Richard McArdle and Walter H. Meyer and published in October 1930 by the U.S. Department of Agriculture. The authors had this to say back about growth in 1930:

> Since the continued existence and prosperity of the lumber industry in the Pacific Northwest is so dependent on the growth that takes place in these still immature stands of Douglas fir and upon areas yet to be logged and reforested, it is important to have definite information concerning the potential yields on these forest lands. The owner of timberland who contemplates raising crops of timber on his land in such succession or alternation as to produce a sustained annual yield of a fixed volume, the investor in growing timber, and the manufacturer looking for a future supply of raw material, all are interested in knowing the growth and yield possibilities of Douglas fir stands.

And none more so than Weyerhaeuser with the world's largest inventory of mature Douglas-fir in private ownership.

For some twenty years Weyerhaeuser research foresters had been measuring growth on 210 sample plots distributed throughout the company's lands. By 1965, they had enough data to compile volume tables, which were put together by Jim King. These proprietary tables showed remarkably higher growth rates than predicted by the 1930 tables of the Forest Service. In addition, soil scientists predicted up to almost 20 percent increase in yields by fertilization. George Staebler's work showed significant increases by periodic thinning. Forest geneticists pointed out what plant breeding had done over the years for agricultural crops like corn and wheat. We could do the same for trees, they claimed, and significantly increase yields of wood. The company abandoned the nineteenth-century measuring system of Scribner board feet and cords and went to cubic measure, which more accurately showed the total volume of fiber. John Downer developed the cunit measure, which represented one hundred cubic feet. All of these developments promised greatly increased yield from growing forests. In effect, the foresters were saying, "give us the money and we can grow wood faster than anyone has ever done before."

In the *Timber Harvest Review* in September 1964, George Weyerhaeuser said:

> We must recognize the difficulties involved in converting an over-aged virgin forest into a balanced, second growth forest. There are real problems involved in setting a course today and adhering to it unswervingly for even ten years, let alone the forty years or more required to grow a merchantable tree. In arriving at a timber plan, we are influenced by our present timber situation, our manufacturing and marketing facilities, the short and long-term projections of supply, demand, and resultant prices, and trends in taxation, to say nothing of the biological factors that affect

> our timber stands with time. As these factors change in the decades ahead and as technological developments bring about new processes and new products, it will be necessary from time to time to adjust our sights and to shift our course. This is the very nature of long-range timber management.

In April 1966, George H. Weyerhaeuser at the age of thirty-nine became president and chief executive officer of the Weyerhaeuser Company. Many of the members of the Board of Directors were pressing for fresh ideas on how to manage the company's vast timberland holdings. At about the same time a financial manager at Crown Zellerbach was saying some things about the financial management of timber assets. Gilbert O. Baker had pushed hard for cubic measure instead of board feet while at Crown, and although he had the support of foresters, he never did convince his senior management at the time. When the CEO of Crown started talking about selling their timberlands to get money for new paper machines, Baker began looking around.

Gil Baker joined the Weyerhaeuser Company on February 1, 1965. He found a fertile field for his ideas—a young aggressive CEO, a staff of bright forest economists, and a research department with some of the top scientists in the industry. He also found himself in the middle of a reorganization of the timberlands department. Although George Weyerhaeuser hired him, he was assigned to work under Harry Morgan.

Since leaving Coos Bay, Harry Morgan had shot up through the company's ranks and now was just under George Weyerhaeuser. If George wanted more intensive management, Harry would give it to him. Ed Heacox, who had built the company's forestry department over the years and had been responsible for the research center, did not get along with Harry Morgan and was soon to retire.

In 1987, long after he retired, Gil Baker reminisced about his relations with Ed Heacox. "Ed was very, very good to me . . . he gave me some counsel . . . and he knew the way things worked." Baker went on to say, "But I also remember him saying to me in his fatherly fashion, 'You know, Gil, what you are doing is immoral. You are saying that you can go out and cut a lot of wood out of this forest and that there will still be forests coming along. You are overstating the case and it's not right.'" Gil believed strongly that if we did everything in the plan it would work, and if we didn't get it all done we would change the plan, just as business plans change when market conditions change. Ed took Gil to the research center and showed him the new yield tables that were almost completed and the intensive Douglas-fir management system that George

George H. Weyerhaeuser, Sr. Courtesy Forest History Society Archives.

Staebler and his staff had put together. Gil realized that all of this was basic to any high yield forestry concept, and Ed had been responsible for it.

Baker and his two economists, Wes Ricard and Phil Woolwine, talked at length about how to get to a target forest with given yield tables and unbalanced age classes and how to handle it financially. To run their projections they needed computers, and Jack Bandel in the inventory department had several computer whizzes including Dick Reagan who could run projections out for a century or more. The first runs made for an actual forest were made for the Millicoma.

With the numbers from Gil Baker and Wes Ricard and the science from George Staebler and his staff, the company began developing a bold new timber management program eventually called "High Yield Forestry." Ricard, then the company's manager of forest economics, and Baker were the "fathers" of the concept. In an oral interview with William Lawrence and George Staebler of Weyerhaeuser's Research Center, Ricard described the program:

> So our theory was we want to build something that you will like when you have it . . . think in terms of building rather than harvesting. So really we just ignored what we had in terms of forests and said, 'What would you like to have, because at some time all you're going to have is what you have grown anyway, so you might as well decide what you

> would like to have.' You brought your mind around to the point where you've accepted the fact that you can't save anything. All you can do is focus on what you're going to build. Well, this immediately released all kinds of volume for harvest, because there was no point in giving up the tremendous values out there to bugs just for some notion of conservative rationing.

This was a revolutionary concept especially for some of the older foresters brought up on sustained yield and Bulletin 201. Ed Heacox called it the target forest, but later the public relations people called it high yield forestry. For years some in the industry and especially Wall Street had characterized the conservative Weyerhaeuser Company as the Sleeping Green Giant and the Bank of Tacoma. With this new concept the company was preparing to make some large withdrawals from the bank.

Not all the company's branch managers were that enthusiastic about this new push from Tacoma. Howard Hunt, when he was the manager at Coos Bay asked what were we going to do with all these six-inch trees when the old growth was gone. Later, when he was the manager at Springfield, he wondered what are we going to tell the people when the mills shut down.

Gil Baker had examined the "dip" problem. In an interview with Lawrence and Staebler, he said, "There are dips everywhere because you want to get to the target forest. You want to get to that efficient condition as early as possible." He went on to say that "any forest program that did not have a dip in it as you go from old growth or wherever you are to the optimum situation is not aggressive enough to do the job . . . so you should have dips . . . maybe you are going to shut a mill down for awhile, or whatever. There are a lot of other things going on at the same time."

In October 1966, Harry Morgan presented the program to the Board of Directors for their approval. With charts and tables he illustrated the increased growth, cash flow, and return on investment for each level of intensive management. It was a financial rotation rather than biological. He stated, "We are confident of being able to grow wood fast enough to justify increasing our cut 37 percent above the level recommended in 1964. This appears to make excellent economic sense for the long run. . . . The basic justification for increasing the cut is intensification of our forest management practices. Only by making rapid and imaginative progress toward improved forest management can we increase our cut while maintaining our land stewardship integrity. This intensification will require the investment of large sums in fertilizing, thinning, nurseries, seed orchards, research, and other necessities for growing wood more

Helicopter loading up seed. Photo Dean Higinbotham.

efficiently. These sums can come from the additional revenues generated by increased timber harvests." Morgan went on to say, "I want to emphasize that this is not a risk free program. The principal result of overoptimism would be a reduced cut some years in the future. Our conclusions are predicated on what we already know and reasonable assumptions about research and technological progress. Attaining these goals will require great effort." The board approved the plan and the following year extended the program to the company's western and southern pine forests. Under the plan, the Millicoma's old growth timber would be liquidated in twenty-five years.

Prior to the adoption of the high yield forest plan, most of the regeneration on the Millicoma had been by aerial seeding. Beginning in 1956, several thousand acres were seeded each year. To obtain the large quantities of seed required, Weyerhaeuser began an aggressive cone buying program in Oregon and Washington. In good cone years such as 1961, over 10,000 bushels of cones were purchased at buying stations established at country stores and gas stations scattered throughout the area. All transactions were cash, and Don Borglum of the forestry staff made the rounds with large amounts of money in

Early planting, 1952. Photo Arthur V. Smyth.

his wallet. It was a short season each fall, but many families made it a day in the woods looking for squirrel caches or climbing young trees. It was dirty work, and the cone pickers would come in covered with pitch. The green cones were shipped north to the company's cone plant at Rochester, Washington, where the seed was extracted. In the fall the treated seed was returned to the Millicoma to be sown on the newly harvested areas. The clack, clack, clack of the helicopter's rotors may have startled the browsing elk, but the foresters were thrilled as the seed rained down on the land.

After the adoption of the high yield program, it became evident that seeding was an inefficient method of reforestation. Spacing could not be controlled, and much of the seed would land on stumps, roads, or down wood—places that made germination impossible. A pound of seed sown directly on the land could reforest an acre. If however that pound of seed was sown in a nursery, it could produce over forty thousand seedlings, which could reforest some seventy acres. Also the cone buying program gave no guarantee of the seed source, but if geneticists could produce trees from certified seed, growth would have a tremendous boost. The company shifted to plantation management, and instead of helicopters in the sky, long lines of tree planters slowly moved up the hills swinging their hoes and tamping down Douglas-fir seedlings. A huge

Geert van Rykevorsel, Weyerhaeuser forester, showing off Douglas-fir growth. Courtesy Weyerhaeuser Archives.

Thirty-five years of growth.

tree nursery was built at Rochester, Washington, and seed orchards were established throughout the region. Millions of trees were planted on the Millicoma Forest.

Helicopters did not disappear from the skies above the Millicoma. The highly productive soils that grew Douglas-fir so well also grew many species of brush that in some cases could over top the seedlings. Scientists developed chemicals that killed or suppressed the unwanted herbaceous species. Paul Lauterbach was the company's expert on chemical spraying, and he spent a lot of time on the Millicoma. Helicopters fitted with booms would dart like some giant dragonfly back and forth across potential brush fields laying down a fine mist of chemicals.

Some dark clouds, however, loomed on the horizon as the chemical spraying continued. In 1962, Rachel Carson's book *Silent Spring* aroused much of the public to the impact on the environment of indiscriminate use of chemicals. At a large regional meeting of the Society of American Foresters held in the early 1970s in Coos Bay, Oregon's governor, Robert Straub, as he addressed the foresters was hit square in the face with a cream pie thrown by a young woman belonging to CATS, Citizens Against Toxic Sprays. In addition, a local lawyer tried hard to persuade a Weyerhaeuser employee who was dying of cancer to sue the company. The employee operated the company's spray truck

Coos Bay five-year plantation. Photo Robert Hanson.

that did roadside spraying. The district's congressman James Weaver got involved at the lawyer's urging, but the employee, a heavy smoker, refused to sue right up to the end. The congressman eventually apologized to the company's representative. Several other people charged that the company's chemical spraying was causing miscarriages in pregnant women, but none of these charges were substantiated.

Helicopters were also used to fertilize the young stands. Nitrogen pellets rained down on some of the stands that had reached a certain age and had been pre-commercially thinned. This was indeed tree *farming*. The growth was astounding as some of the Douglas-firs growing along Matson Creek were putting out leaders that exceeded four feet in a year. It may have been farming, but the young plantations were not in neat rows like a cornfield. The planters had to place the trees wherever they found a place between the stumps that would ensure survival, and there was usually fill-in from natural seeding. These young plantations could not be differentiated from a natural stand. Much of the forest was now resembling what it looked like in 1775 at the very beginnings of our nation.

The growth was exceeding the estimates. But loggers quickly moved out the old timber. Logs were streaming down the Millicoma, the South Fork, and out the back end in timber sales and logging contracts in the Roseburg area. In 1963, a new plant joined the sawmill and particle board plant on the Coos Bay waterfront. The first automated Douglas-fir plywood plant, a technological marvel, attracted visitors from around the world. It could produce plywood panels at an astonishing rate. However, it couldn't seem to produce a profit. To increase profitability, foresters developed log forecasting systems so that the plants downtown would have some idea of the size and grade of the logs that would be coming out of the forest.

Howard Hunt at one time referred to the Coos Bay operations as the "Coos Bay Academy," as more and more of its managers moved up the corporate ladder. Harry Morgan left in 1957 to become the manager at Snoqualmie Falls. Eventually he became the executive vice president of the company and was the chief corporate sponsor for the high yield forest program. In 1963, Herm Sommer went to Twin Harbors. In 1964, Clyde Kalahan moved to Springfield to take over as branch manager. In 1966, Howard Hunt moved up to Tacoma and was succeeded by Oscar Weed. Just as Hunt was different than Art Karlen, so was Oscar Weed different than Howard Hunt. Hunt was an engineer and a numbers man; Oscar was a salesman and a people man. Oscar liked people, and people liked him. His enthusiasm could be infectious.

Oscar Weed. Courtesy Weyerhaeuser Archives.

During the 1970s and 1980s many managers, loggers, foresters, and engineers left their mark on the Millicoma—Ken Schaefer, Skip Dunlap, Don Wickendoll, Paul Shook, Jack Wolf, Steve Conway, Ted Nelson, Jim Rombach, and many others. Foresters included Don Gingery, Fred Dawson, Jim Lavan, Pete Belluschi, Bob Wiggins, Dick Lamb, Jim Larson, Leverett Curtis, Bob Patterson, Ralph Sweet, Harry Spencer, Jack Palmquest, Floyd Page, Don Borglum, Jack Winjum, and R. A. "Al" Petty, who outlasted them all working for thirty years on the Millicoma. Don Wickendoll, who succeeded Skip Dunlap as logging superintendent at Allegany in 1968 and continued that job until the Allegany terminal was shut down, became the longest tenured logging boss at Allegany.

Another rising star in the Coos Bay operation was Jack Wolff. Jack had made a name at the company's Klamath Falls operations, the only western pine holdings the company owned. He succeeded Phil Hogan as woods manager reporting to Oscar Weed. Coming from the flat pine lands, where he was woods manager, Wolff was struck by how steep and broken up the country was.

One of Wolff's first jobs was to convince Rex Allison to accept the new job of forest engineer, where he could oversee the foresters. But Rex preferred the job of woods manager. Rex finally accepted the offer, and Paul Shook became the superintendent at Dellwood. Wolff was told by Oscar Weed that high yield forestry was for real and that his job was to see that it got implemented. Rex surprised everyone and supported the foresters all the way.

The forest was changing too. The site of Dan Mattson's homestead cabin was now surrounded by stumps, but the stumps were hidden by the thick growth of young Douglas-firs. The creek that bears his name still wound through the flat and plunged over the falls on its way to the sea. But society was changing too, and events that were occurring three thousand miles away would affect our forest just as did fire, wind, beetles, and flood. The only constant were the winter rains, summer sun, and another ring of wood on the trees marching up the hillsides.

References

Annual Forestry Reports: Millicoma Forest. Weyerhaeuser Company Archives, Tacoma, Washington.

McArdle, Richard E., and Walter H. Meyer. "The Yield of Douglas Fir In The Pacific Northwest." *Tech. Bulletin 201*. (1930).

Morgan, Harry E., Jr. "The Fir Target Forest." Presentation to the Board of Directors. Weyerhaeuser Company, 1966.

Oral interviews, 1987. William Lawrence, George Staebler, Wes Ricard, Gilbert O. Baker. Weyerhaeuser Archives. Tacoma, Washington.

"We are concerned about air, water, global warming, owls, woodpeckers, ecosystems, ancient forests, tropical deforestation, our natural world, the sustainability of our natural resources."

—A.V. Smyth Address, University of Michigan, February 22nd, 1990

12 / A Changing Society

As Aldo Leopold pointed out, "From the earliest times one of the principle criteria of civilization has been the ability to conquer the wilderness and convert it to economic use. To deny this criterion would be to deny history." From our nation's very beginnings the great American forest played a vital role in the lives of its citizens. To the early colonists the forest was a fearsome thing that had to be cleared to provide food and shelter for their growing families. As the country grew the forests were cleared at a prodigious rate to provide food for the growing army of immigrants coming into the promised land. The trees were used to construct the houses, barns, and fences of farms across the country. Wood was building railroads, cities, churches, and schools. And it provided the fuel to heat the homes and fire the forges. America was hell-bent on building a country—with wood.

The great American forest was also an inspiration for the country's poets, artists, and philosophers. Henry David Thoreau, John Muir, Ralph Waldo Emerson, and William Cullen Bryant all expressed their reverence for the natural world and the wildness of the American landscape. Later in the twentieth century, Leopold, undoubtedly the most eloquent forester of our time, asked, "Shall we now exterminate this thing that made us American?" The debate

Teddy Roosevelt. Courtesy Forest History Society Archives.

about the nation's policies regarding its forests has waxed and waned for well over a century, and it touched the Millicoma as well.

In 1905, Dan Mattson had built his homestead cabin above the falls. The forest around him was now about 140 years old, and the huge trees dwarfed his little clearing. Three thousand miles away, the president of the United States, Theodore Roosevelt, and his forester, Gifford Pinchot, had organized probably the most important meeting on forest issues ever held in this country. The Second American Forest Congress marked a radical change in the country's land and timber policies. During this period, millions of acres of the public domain were withdrawn from entry, some of the most abused land laws were repealed, the United States Forest Service was founded, and the National Forest system came into being.

Pinchot invited Frederick Weyerhaeuser to the meeting on behalf of the president. Frederick's son, Frederick E., wrote back that his father was ill and could not attend. Pinchot then invited the young Weyerhaeuser and asked him to present a paper. Years later in a letter to his son Davis, Weyerhaeuser told about his experiences at the Congress. "In my youthful enthusiasm I agreed to, and making rather full use of my father's name and influence, at Pinchot's request I urged many prominent lumber manufacturers and timber owners of the country to accept an invitation to attend the Congress, which invitation Pinchot let it be known came from the President."

Roosevelt opened the Congress with words of welcome, and the young lumber scion, F. E. Weyerhaeuser, was an early speaker on the program. He told the Congress, "Practical forestry ought to be of more interest and importance to lumbermen than to any other class of man. Forests will reproduce themselves if given a fair chance but there are three great obstacles which must be reckoned with in the profitable reproduction of timber; time, fire and taxes." He also said, "On the Pacific Coast also the climate is suitable for the steady and rapid growth of excellent timber. At the present time values there are too low to insure any profit in conservative forestry, but a few years will undoubtedly bring about very different conditions." As to forestry, Weyerhaeuser said, "Forestry is a new idea to us and we have given little thought to the future."

After the preliminary remarks, Roosevelt shocked Weyerhaeuser and his industry associates by turning to the over one hundred lumbermen seated on the stage behind him. Barely recognizing the audience and completely ignoring the remarks prepared for him by Pinchot, the president lit into the startled dignitaries. In his letter to his son, Weyerhaeuser described it vividly. "He shook his fist, his teeth gleamed, he called them skinners of the soil, despoilers of the national heritage, and shouted many other insulting and characteristically Rooseveltian phrases, all intended for the newspapers, and thereby for his own glorification. He received it in full measure. Incidentally, I received the reasonable abuse of most of my friends, whom I urged to come to the conference." Weyerhaeuser told his son Davis, "Without much question, his was the greatest personality of his generation, but I never liked him. His tremendous egotism and desire for self exploitation, regardless of fairness to others, always disturbed me, but why put the magnifying glass on the weaknesses of great men."

The meeting, and subsequent policy changes, resulted from the public's dismay over the "cut out and get out" practices that had resulted in thousands of acres of burned and denuded pine lands in New England and the Lake States. The conservation movement had many voices reaching back to the 1840s, but Teddy Roosevelt and his forester, Gifford Pinchot, gave it national recognition.

For many decades there was little or no reforestation of cut-over private forest land. No prudent landowner or investor would plant trees if they burned up in a few years. F. E. Weyerhaeuser had also pointed out that along with fire, taxes were another obstacle to forestry on private lands. During the 1930s millions of acres of cut-over lands were being abandoned to the tax collector.

Foresters were urging the Weyerhaeusers to hold on to their cut-over land, protect it, and reforest it. Dave Mason, a prominent consulting forester, pushed his ideas for sustained yield forestry and convinced F. E. Weyerhaeuser that planned forestry was the only real way out of the industry's difficulties. He met with George Long in 1927 and with the executive committee of the Board of Directors in St. Paul in 1930. While the top management was dedicated to this idea of sustained yield, getting the logging managers in the 1930s to put it into practice was a different matter. Ed Heacox, one of the first foresters hired by the company, was assigned the job of selling forestry to the logging superintendents. It was a tough sell in the 1930s. For example, when Heacox suggested to logging boss Ronald McDonald that he leave some seed trees at the Vail operations in Washington, McDonald snorted about, "those damned Roosevelt seed trees." This Roosevelt was FDR. But in 1941, at the Clemons operations in Grays Harbor County, Washington, an idea was hatched that subsequently swept the country.

The tree farm program, which began in June of 1941 at Montesano, Washington, was an attempt to respond to the public's concerns about forest management. A very long-range business decision, much of it was based on faith. Industry was justifiably proud of its accomplishments as were the small landowners of their newly certified tree farms. But some people were suspicious. Lyle Watts, Chief of the Forest Service in the early 1940s, claimed that the tree farm program was nothing but a publicity ploy to forestall government regulation. David Brower, then with the Sierra Club, sneered that a tree farm was "one sign and a million trees cut down." But the program had many supporters. Chief Ed Cliff later lauded the program and called industry efforts important. Samuel Trask Dana, then the dean at the University of Michigan, called the program enlightened self-interest. Henry Clepper of the Society of American Foresters pointed out that it was an independent American phenomenon developed without government support.

Forestry, from the beginnings of the tree farm movement in the Willapa Hills of Grays Harbor to high yield forestry on the Millicoma, had made quantum leaps. But the world was changing. American society, which began as an agrarian, rural society of small towns dependent on the land, was now an urban society of teeming cities and affluent suburbs far removed from the land. The distance between the affluent commuter heading into New York City from the suburbs on a commuter train, and the small knot of loggers waiting for the crummie on Front Street in Coos Bay, Oregon, was far more than 3,000 miles in geography. It was becoming a widening chasm of different values, perceptions, experiences. The forester's world was about to be turned upside down.

References

American Forestry Association. *Proceedings of The American Forest Congress.* Washington, D.C.: H. M. Suter Publishing Co., 1905.

Flader, Susan L., and J. Baird Callicott, eds. *The River of God and Other Essays.* By Aldo Leopold. Madision: University of Wisconsin Press, 1991.

Hidy, Ralph W., Frank Ernst Hill, and Allan Nevins. *Timber and Men.* New York: The Macmillan Company, 1963.

Smyth, Arthur V. "The First Tree Farm." *Virginia Forests* 49, no. 2 (1993).

Weyerhaeuser, Frederick E. Personal letter to C. Davis Weyerhaeuser, 1932.

"Let them have dominion over the fish of the sea, and over the fowl of the air, and over the cattle and over every creeping thing that creepeth upon the earth."

—Genesis 1:26

13 / More Than Trees

The Millicoma Forest was the home of an abundant variety of fish and game that, like the trees themselves, were looked upon as commodities or, in the case of predators, as a threat to domesticated stock. From their beginnings, organisms, both large and small, have evolved and changed as their habitat changed. And from the very beginnings humans have manipulated their habitat, the natives with fire and bow, the white man with ax, rifle, plow, and bulldozer.

Roosevelt elk were and still are the largest non-human inhabitant of the Millicoma Forest. Long before the Millicoma became a tree farm, the settlers slaughtered the elk for their hides, their teeth, and their meat. In the 1880s, George Gould packed his family into what became known as Elkhorn Ranch up the West Fork of the Millicoma. He and his family were market hunters, and they killed elk by the hundreds, chiefly for their hides, which were salted down and taken by trail out to the Umpqua. In one three-day period they reportedly killed eighty-five elk. At one time the Indians prized the two "canine" teeth that the elk had in their jaws as currency, but the big market were the members of the Benevolent Order of the Elks.

Kentuck Thomas and James Jordan were also market hunters during this period, and according to testimony cited by Lionel Youst, Jordan would take orders from sailing ships in the harbor for elk meat and go into the forest to fill the orders. He delivered as many as six elk to each vessel. In 1891, one thousand elk were killed, mostly for skins, in Coos County. Almost all of the hunting took place in the 1868 burn, which had resulted in vast open areas covered with grass. The Millicoma Forest, still inaccessible in the 1880s, was little hunted then and must have provided a refuge for the elk. By the late 1880s, the elk herds were so decimated, as a result of market hunting and extreme pressure on the herds in the rest of the state, that the state legislature prohibited the killing of any more elk in the state. The hunting season was not reestablished for elk until 1940. The Oregon State Department of Fish and Wildlife stated in 1987 that "it's easy to forget the incredible success story. Forty-three thousand elk wintered over in 1985 on prime Roosevelt elk ranges where early in this century extinction seemed imminent. Distant predictions in future timber output models are gloomy for elk, but for now, miles of habitat in, or approaching optimum conditions, beckon."

The black-tailed deer were, next to the elk, the most numerous of the native herbivores, and here again they were not only a source of food but also a cash crop for some. During the Depression, Bill Leaton, a Glenn Creek homesteader, killed 103 deer one year. Much of the meat was "jerked" and the dried meat sold to "doctors and lawyers" in town. He sold the hides to the Marshfield Bargain House. Before the logging began on the Millicoma, deer were not all that plentiful, but as the country was opened up the deer population soared along with the elk. In 1967, hunters harvested 245 deer from the Millicoma.

The fur-bearing mammals of the Millicoma were another source of income for the early settlers. Many trappers (most without a permit) ran trap lines through the Millicoma. Trappers respected each other's territory, but if it was abandoned another trapper could move in. Cle Wilkenson, a long-time Glenn Creek settler, trapped every winter and had a cabin on a tributary of Matson Creek in the 1930s. During the Depression, Wilkenson got a job as a WPA predator trapper. The county paid bounties of fifty dollars for a cougar and five dollars for a lynx or bobcat. With the WPA job, Wilkenson could now trap during the summer months. Marten and mink were the most profitable fur bearers. The season on marten closed statewide between 1937 to 1945 and then again from 1947 to 1950. Between 1961 and 1995, only thirteen marten had been trapped in Coos County and none since 1988. In 1994, the marten was listed as "state critical" by the state of Oregon.

Early newspaper reports were filled with stories and photos of cougar and bear kills. In one year near the turn of the century, 135 cougars were killed in

Steelhead trout from Millicoma, 1954. Photo Arthur V. Smyth.

Coos County. Another story told of ten bear being killed in one day. Bobcats and lynx seemed fairly common, but coyotes were rarely mentioned. Bounties were placed on cougar and other predators in Oregon as early as 1843. From 1912 until 1963, 502 cougar were turned in for bounty in Coos County. By 1961, it was estimated that only 200 cougar remained in Oregon and unless protected the cougar would be eliminated from the state by the early 1970s. The Game Commission repealed the bounty system in 1961 and classified the cougar as a game animal in 1967. The state closed the season in 1968 and tightly regulated the taking of cougar. The Oregon Department of Fish and Wildlife estimated in 1993 that Oregon had a cougar population of 2,500, and evidence indicates that the population is increasing.

Salmon and steelhead trout were plentiful in the Coos and Millicoma Rivers below the falls and were a major source of food in the fall for the native tribes before the coming of the Europeans. The East and West Forks of the Millicoma contained thriving populations of native cutthroat trout as did Lake Creek and all the tributaries of the South Fork. The early splash damming on the East Fork must have degraded the aquatic habitat as did the splash damming on the South Fork. According to the interviews Youst conducted with Alice Wilkenson and others who had lived in the Glenn Creek valley above the falls, when Joe Schapers and his brothers first homesteaded above the falls, no fish occupied the stream. They were blocked by the three-hundred-foot falls. Sometime around the turn of the century, Joe and his brother Jerd carried cutthroat trout in two five-gallon coal oil cans over the falls and planted them in Glenn Creek where they flourished. To keep the fish alive the Schapers would pour the fish from one can into the other to aerate the water. According to Warren Browning, also interviewed by Youst, the Schapers also planted eastern brook in Matson Creek above the Matson Creek falls. The Eastern brook trout probably came from Frank Smith's hatchery on the South Fork. According to fisheries biologists, Matson Creek in Coos County is the only stream on the west side of the Coast Range where Eastern brook trout were successfully planted. According to legend, during the Depression, fishermen who packed in over the falls took out trout by the barrel.

When the loggers moved into the Millicoma in 1950, they encountered the other inhabitants of the forest. Elk were everywhere, and big bulls were commonly sighted. Marten would come up to the logging crews eating lunch on the landings and run away with scraps the loggers would throw them. Ovie Coleman brought a bear cub down to the filing shack in his pickup. The cub had come down with a falling tree. After showing it off, he returned it to the woods. Jim Lavan working in Cedar Creek had a standoff with a huge bear feeding on a deer carcass. It was a tense few minutes before the bear retreated. Dick Lamb coming down the 2000 line was startled by a cougar bounding across the road. Bobcats were frequently sighted along the logging roads, and coyotes were becoming common. A family of beaver was busy damming up the culverts on Conklin Creek. And the Eastern brook trout in Matson Creek—were they still there?

Lou Hoelscher in the early 1950s lived at the Allegany camp when he was assistant bull buck. He was an avid fisherman as was I. Lou had been bragging to me about the secret fishing spot he had where he caught large Eastern brook trout on a fly. He refused to reveal his secret but promised that he would tell me where it was, when and if he ever left Coos Bay. Lou was promoted to Tacoma where he became safety director and training administrator. The moving truck was in the Hoelscher's driveway when I ran up to him. "Lou, you haven't told me where your secret fishing spot is," I pleaded. "Okay, Smitty, I am leaving, but you must promise to keep it a secret." It was then that he told me. "It's Matson Creek. Just go up the 3000 line, anywhere along it. Believe me, you will be surprised." I was dumbfounded. In the 1950s, the 3000 line was the main logging road into the operating area. Hundreds of trucks came roaring down the road every working day. Matson Creek, which paralleled the road was a relatively small stream with clearcuts on each side. Could this be

(clockwise from top left)

Gretchen Smyth with Eastern brook from Matson Creek. Photo Arthur V. Smyth.

Eastern brook from Matson Creek. Photo Arthur V. Smyth.

Pools created on East Fork Millicoma for salmon spawning.

possible, I thought? At that time I was completely unaware of the Schapfers or Frank Smith.

It was a Saturday. There would be no hauling on the 3000 line, and there would be no one in the woods early in the morning when I drove to Hoelscher's secret fishing Nirvana. I strung my rod, tied on a streamer, and beat my way through the streamside brush. Stepping over logging slash and wending my way through the stumps, my heart was pounding as I stood on the bank of the fast-moving stream. I flipped my fly into a riffle and watched it disappear under a log. Then the rod bent. The line shot out. I had a fish—a big one. In my excitement I am afraid I "horsed" it out, but everything held together. There on the bank lay a shimmering, sixteen-inch speckled beauty, my first ever Eastern brook trout. In an hour I had six trout from twelve to sixteen inches in my creel. Yes, Joe Schaper's trout were still there.

Deep in the forest there was one "fowl of the air" that was unnoticed and never mentioned in the stories of the early pioneers. This small, dark-eyed creature with its appetite for voles and a voice that resembled a dog's bark would have more influence on the forests of the Northwest and the Millicoma than wind, beetles, or fire. The Northern spotted owl was its name, and to some it was a name that would live in infamy. But that was yet to come.

At the beginning of the twentieth century, forests, and even many of the creatures that lived in the forest were treated as commodities. As we neared the beginnings of the twenty-first century, the American people viewed them not as commodities but as precious resources. It was no longer *fish and game*. It was wildlife, which included "every creeping thing that creepeth upon the earth."

References

Harper, James A. et al. *Ecology and Management of Roosevelt Elk in Oregon*. Portland: Oregon Department of Fish and Wildlife, 1987.

Mahaffy, Charlotte L. *Coos River Echoes*. Portland, Ore.: Interstate Press, 1965.

Peterson, Emil R. *A Century of Coos and Curry*. Portland, Ore.: Binford and Motts, 1952.

Rosen, Rudolph. Letter from Oregon Department of Fish and Wildlife to Arthur V. Smyth. *Trapping Statistics and History of Cougar Management*, 1996.

Youst, Lionel. *Above the Falls: An Oral and Folk History of Upper Glenn Creek, Coos County, Oregon*. Coos Bay, Ore.: South Coast Printing, 1992.

"The supply of logs during the next thirty years will come largely from old growth timber with a gradually increasing amount from new growth. Beginning about 1980 the yield from old growth will diminish rapidly and the yield from new growth will be proportionately increased." —Weyerhaeuser Press Release, 1950.

14 / The Transition

During the 1960s and throughout the '70s and '80s, the Millicoma Forest saw the rapid harvest of the old growth stand. The first decade of logging on the Allegany side of the forest, the 1950s, produced a little over one billion board feet of timber. All of it came down the now black-topped mainline road into the Allegany terminal.

With the opening of the Dellwood side in 1961, production from the forest soared. It only took five years to produce the next billion feet. Some of this volume was made up of timber sales to other markets; some of it into Roseburg. By the end of the second decade, the 1960s, the forest had been depleted by over 2.8 billion feet. During the 1970s, which saw some boom times for the industry, the Millicoma produced 4.3 billion feet of logs, primarily Douglas-fir. And in the 1980s up to the day the big mill closed in March of 1989, an additional three billion feet came off the forest.

From that summer day in 1950 when the whine of a power saw broke the silence of the Millicoma Forest to the day when the last log went through the big sawmill on Coos Bay's waterfront, our forest had provided over ten billion feet of timber. More than a production statistic, that figure also represented the dreams and hopes of thousands of people over almost four decades. It provided for homes, college educations, and retirements for countless families. Many millions of dollars went to the local governments for schools and roads. It was a way of life for thousands of workers who went up into the forest to build the roads, fall the timber, and drive the trucks that brought those millions of logs out of the remote recesses of the Millicoma Forest. And people thousands of miles away from this Oregon forest may have been living in homes or worshipping in churches or attending schools built with lumber from these logs. For the past forty years the Millicoma Forest had been a powerful economic engine. As we entered the last decade of this century, many people in the community were worried that perhaps the engine had run out of gas.

Our forest was now profoundly changed. Communities had changed, and people had changed. One thing that had not changed through the 1960s and 1970s was the company's commitment to high yield forestry. The forest was now laced with thousands of miles of roads stretching from Bear Mountain overlooking the Umpqua Valley in T27S R8W to the second growth stands tributary to the sloughs opposite Coos Bay. Planting behind the logging was now just as big a business as harvesting. Reforestation companies supplied the labor for this business. Chiefly Hispanics, most are now citizens and live in the area. Some have been planting trees on the Millicoma for twenty years.

In 1976, Don Borglum, a member of the forestry staff, said, "When you plant nearly two million trees which we did in four and a half months, it takes a lot of doing. We had as many as 240 people planting trees." In 1986, 2.4 million seedlings were planted. From the first planting in 1949 to 1986, 47 million trees had been planted on the Millicoma. The trees being planted in 1996 are far superior to what were planted in 1949. Today the trees are grown from seed that comes from company seed orchards that produce genetically improved seed from known genetic families.

Before planting, the site is treated with chemical herbicides. With the competition from the herbaceous vegetation eliminated, the seedlings, according to forester Jim Clarke, "just blow out of the ground." In three years the average tree is five and a half feet tall. Our forest was now, for the most part, a human-made plantation forest growing wood at a prodigious rate.

Allegany was changing too. In 1961, the apartments were sawed in two, loaded onto short logging trucks, and hauled over the 5000 line to Dellwood where they served as offices. In the winter of 1967, the company houses were put up for sale. A condition of the sale was that they would have to be moved out of Allegany because the site was needed for a log yard. John Kruse, who lived on the East Fork just above Bridge 1, bid on several and bought them for $350.00 each. He moved them onto his property where he used them as rentals. Others went to various other locations. House #2, the only red house in camp and our home for four years, is now lost somewhere in Coos County.

The downtown also changed. The company acquired Menasha's paperboard mill in 1981 and finally gave up on the plywood mill, shutting it down permanently in 1984. The following year it was sold to Sun Plywood. But much more significant changes lay ahead. The transition was not yet completed.

References

Borglum, Donald. Oral Interview, 1976. Weyerhaeuser Archives. Tacoma, Washington.

Clark, James. Oral interview with author, Coos Bay, Oregon, December 1995.

Forestry Annual Reports. Weyerhaeuser Company, North Bend, Oregon.

Kruse, John. Interview with author, April 1996.

"There was a twinge when I watched the last raft leave Allegany." —Don Wickendoll

15 / The Last Raft

On January 4, 1989, Don Wickendoll, the logging superintendent at Allegany since 1968, watched the last raft go down the Millicoma River. Another chapter in the history of the Millicoma had ended. No longer would the tugboats wend their way past the dairy farms, towing the logs from Millicoma and Lake Creek to the mill on the waterfront. The river that in years past had provided transportation for school children and milk cans had now ceased being a carrier for logs.

Ed Boren brought down the last load of logs into Allegany. Houston Weeks, in his boom boat, pushed the logs into the last raft. After almost forty years the log dump at Allegany became silent. For all those years the siren at the unloading dock would sound before the logs from each truck would be catapulted into the river. In a typical year, over 20,000 truck loads went over the brow log at the Allegany terminal. Almost six billion feet of logs had come through the Allegany terminal. Now what logs came down the blacktop mainline would go to town down the county road.

Rafting down the South Fork of the Coos from Dellwood was also discontinued. Logs would be unloaded to a dry land storage yard and transported to the mills by truck down the county road. Well over a century of river transportation of logs had ended. The hand loggers in the early 1870s used the winter freshets. The splash dam loggers beginning in the 1880s and the Weyerhaeusers beginning in 1950 all had depended on the river to get their logs out of the forest. Dellwood was now the center of activity with shops and woods offices.

In Allegany, the shop, woods office, and forestry building were now abandoned. Some of the large timbers and lumber from the shop buildings were salvaged or sold. A few months after the last raft had gone down the Millicoma, Don Wickendoll, Ralph Sweet, and Don Wilbur supervised the burning of what remained of the Allegany terminal. As Royce Cornelius's hose tower collapsed into the flames, the Allegany terminal passed into history.

Bob Conklin's first logging plan had predicted that for the first twenty-five years the logging would take place in the Millicoma and Lake Creek drainages. The Allegany terminal had lasted longer than that by several decades. Allegany had been the site of other logging camps long before Weyerhaeuser arrived. In 1890, Henry Laird built a railroad up what is now known as Woodruff Creek. Today the 1001 line runs up along the stream, and the timber that came in after Laird's logging is now harvested by Weyerhaeuser.

On Thursday, March 4, 1989, at 1:24 P.M. the last log went through the huge band saw in Weyerhaeuser's big mill on North Bend's waterfront. The next day the mill fell silent. Although the event was traumatic to many in the community, it was not a surprise. In November 1988, Jack Taylor, manager of the Coos Bay operations, announced that the second shift in the mill would be curtailed effective January 9, 1989. The move would mean laying off fifty employees. Taylor pointed out, "We're facing an overall change in the nature of our raw material base. It's a transition from an old-growth economy to a second-growth economy." In reporting on the lay-off of the second shift, *The* [Coos Bay] *World* stated, "Company officials have said for some time that the North Bend mill would face closure or retooling at the end of this decade." Taylor confirmed the prediction but said they were still studying the options and no decision had been made as of that time.

By the beginning of 1989, it became apparent that the big, old mill designed for old growth logs could not be refitted for the type of wood now coming from the forest. On January 4, 1989, the company announced the closure of the mill effective March 5. The history of the forest products industry is replete with mill closings—from Electric Mills, Mississippi, to Muskegon, Michigan. Those abandoned mills of the last century left in their wake the smoking ruins of cutover forests. But in the hills back of the Coos Bay mill, a new forest was thriving. The thirty-eight-year old mill was ill equipped to use the trees from the new forest, and it was impractical to try to convert it. It was time for a fresh start.

Reactions to the announcement of the closure were mixed. The union representative said, "Some people will be angry, others are relieved." Some blamed log exports. At the same time the company announced the closing of the North Bend mill, it also reported that it had purchased the idle cedar mill from Menasha Corporation. This mill, although relatively new, had closed down because it could no longer get cedar logs. Weyerhaeuser planned to redesign the mill to cut high quality, metric-sized lumber for the Japanese export market. Taylor stated, "We view the CB cedar facility as our 'Mill of the 90s' With some modifications we can efficiently produce high quality, high value finished lumber products to serve the emerging Japanese lumber market." The company ceased its log exports from Coos Bay in 1990.

Because the new mill, which was scheduled to open in April, would employ only seventy-five workers and the old mill had two hundred, there was spirited bidding for the new jobs. Although the company said that former employees would have first crack at the jobs, it was picking the "best of the crop." Some of the laid-off workers enrolled at the local Southwest Oregon Community College to try to develop new skills. Many of the older workers retired, and some left the community. North Bend city officials voiced concern about their budgets and the local economy. Tax assessment for the mill equipment and land amounted to over nine million dollars. The new, smaller mill was located in Eastside, but it was still the biggest employer in the Coos Bay area.

The last day of the mill saw a lot of handshaking, reminiscing, and saying good-byes. Doyle Dickey had worked in the shipping department for thirty-eight years and had helped ship lumber all over the world. Planning to retire, he found it hard to believe that there would not be a log cut in North Bend. Thomas Munyon had worked in the mill since 1960 and managed to send seven of his eight children to college. John Peak came to work in the mill at age nineteen, right out of military service, and worked twenty-nine years as a lumber grader. Cheryl Howell worked at the mill for eleven years and was delighted that she and her father were selected for the new mill. But she loved the big mill: "It's huge; it's monstrous. My kids will never know what a real huge sawmill is like . . . the sound and the smell . . . the pitch and the wood. That is the most unmistakable smell in the world. I'll miss it."

Called CBX, the new mill was one of the few in the industry devoted to the export of finished lumber. It was highly productive, some claiming because of a new look at labor relations. There was a good deal more participatory management, and the union even permitted merit raises. One manager compared the morale and productivity of the new mill with the old by recalling, "We had 800 to 900 people here in the old plywood and lumber mills and we'd get 60 grievances a month on everything from demanding more money for pushing a new button to fights and drinking." He contined, "We turned that around by looking for people who could focus on a positive attitude. And we are communicating better with them . . . this is the best crew I have ever managed."

According to Jack Taylor, by starting over with better equipment and a stronger commitment to quality, the CBX mill received rave reviews on everything from sawing accuracy to safety and cleanliness. It was an example of the changing technology, changing markets, and changing economics that is sweeping across the Pacific Northwest. But something else was sweeping across the Northwest. It had wings, and it threatened the mill itself.

References

Gerson, Greg. "Weyco Buys Mill, to Close NB." *The* [Coos Bay] *World,* January 4, 1989.

Griffith, John. "Last Logs Cut at Weyco." *The* [Coos Bay] *World,* March 3, 1989.

———. "Last Load of Logs Leaves Allegany Today." *The* [Coos Bay] *World,* January 4, 1989.

"Weyerhaeuser: A Sleeping Giant Awakens." *The Sunday Oregonian.* December 29, 1991.

Wickendoll, Don, Don Wilbur, and Ralph Sweet. Telephone interview by author, 1996.

Weyerhaeuser Company News Release. "Coos Bay Shift Curtailment Announced Today." November 10, 1988.

Weyerhaeuser Company Oregon Forest Products Division Update. North Bend, Oregon. November 10, 1988.

Woods Newsletter, Weyerhaeuser Co. "The Last of Allegany." Weyerhaeuser Company, North Bend, Oregon. January 17, 1989.

"To provide a means whereby the ecosystems upon which endangered and threatened species depend may be conserved and to provide a program for the conservation of such endangered species and threatened species."

—Sec 2(5)b,16 USC. 1531

16 / The Owl and the Millicoma

Don Christensen, from a pioneer Coos Bay family, called me around Christmas a few years back. His high-pitched voice almost rattled the receiver: "Smitty, you won't believe what's going on back here. They are bringing in six-inch pecker poles while the old growth is rotting away in the hills, and now some damn fool is saying that you can't cut anything for fifty miles inland from the coast because of another damn bird!" He was referring to yet another endangered species, the marbled murrelet.

By 1990, although most of the old growth was gone in the Millicoma, many six-inch peckerpoles were sprouting. But even this activity was being threatened by a two-pound, dark-eyed owl that the U.S. government had decreed, "thou shall not take." "Taking" also meant disturbing the habitat.

The early planners for the Millicoma Forest were certainly aware of the risks from fire, wind, and bugs. They were also aware of the risks of taxes, but an owl? Of course the issue was more than the owl. The story of how this law was enacted, interpreted, and enforced is one of the most intriguing chapters in Millicoma history.

Charles Mann and Mark Plummer, in their book *Noah's Choice,* date the "roots of the political effort to protect biodiversity" to October 1956 when

Roy Ericksen, a federal wildlife biologist, met in Washington with a group concerned about the future of the whooping crane. The wild whooping crane population was down to twenty-four, and Robert Allen, an Audubon Society ornithologist, was passionately committed to saving this magnificent bird from extinction.

The Fish and Wildlife Service had established a Committee on Rare and Endangered Wildlife Species in 1964. According to Mann and Plummer, by August 1964 a preliminary draft had identified thirty-six birds, fifteen mammals, six fish, and five reptiles and amphibians as threatened or endangered in the United States. By 1968, when the list was first officially published, it had grown to 142 species. The Endangered Species Act of 1966 was passed with practically no opposition. This law gave the Fish and Wildlife Service the authority to spend up to $15 million to buy habitat for species facing extinction. It also directed the Agriculture, Defense, and Interior Departments to preserve the habitat on the lands under their jurisdiction insofar as is practicable and consistent with their primary purpose. Needless to say the agencies used "practicable" whenever there was a conflict with their mission and biodiversity.

The Endangered Species Act was written by E. U. Curtis "Buff" Bohlen, a man dedicated to preserving endangered species, and with impeccable Republican credentials, used his position as Undersecretary of Interior to advance his cause. He took it on himself to write the new bill, which the president sent to Congress in February 1972.

Frank Potter, the counsel for the House Merchant Marine and Fisheries Committee, and Lee Talbot, the senior scientist for the newly created Council of Environmental Quality were Bohlen's "primary movers." Potter thought Bohlen's bill too weak, and he and Talbot worked to toughen it up. Their goal was to get rid of the word "practicable" and to leave as little wiggle room as possible. They rewrote the preamble so that it now gave the federal government the mandate to protect the habitat. Mann and Plummer quote Potter as saying, "That's where we really stuck it to them."

As the bill advanced through the committees and on to the floor of the House and Senate, the conspirators kept their heads down. Few, if any, members of Congress had the foggiest idea of what they were doing. In the conference committee the last vestiges of "practicable" were removed. The Senate voted for the bill unanimously, and the House only recorded four "nay"s. President Nixon signed the bill on December 28, 1973.

Congress had given enormous power to a relatively obscure federal agency, the Fish and Wildlife Service. Not only did Congress have no idea of the sig-

Spotted owl habitat in second growth stand.

nificance of what they had created, but the forest industry lobbyists had missed it almost entirely. It was a sardine-sized fish called a snail darter that suddenly woke up Congress to the monster they had created. For some years a fight continued to brew over the Tennessee Valley Authority's (TVA) plans to build Tellico Dam on the Little Tennessee River. After years of delays, construction had started again in 1973. That summer a University of Tennessee ichthyologist, David Etnier, was snorkeling downstream from the partially completed dam when he spotted and collected a fish new to him—the snail darter. Etnier had been an opponent of the dam and had testified against it.

Hiram G. Hill Jr., a law student at the university, and his professor, Zygmut J. B. Plater, also opponents to the dam, convinced the Fish and Wildlife Service to list the darter as endangered, which they ultimately did. They notified the TVA that the dam would destroy the snail darter's habitat and that construction would have to stop. TVA, believing that no one in his or her right mind would stop construction of a multimillion-dollar dam three quarters away from completion for an ugly little fish, ignored the order.

The TVA had misjudged the passion of the young University of Tennessee law professor and his graduate student. These energetic activists and others mounted a campaign to take on the powerful TVA. After forming the Tennessee Endangered Species Committee, which now included an assistant dean at the law school, Donald S. Cohen, the group went to court. *TVA* v. *Hill* finally reached the Supreme Court, where Griffin Bell, the Attorney General of the United States, argued for the TVA. To his dismay, in June 1978, the court, by a vote of six to three, favored the snail darter. Howard Baker, the senior senator and minority leader from Tennessee, managed an amendment to the act that formed the Committee on Rare and Endangered Wildlife Species. This committee could be convened when all other efforts failed to resolve a conflict. The committee was authorized to permit a species extinction when the benefits of the project clearly outweighed any other alternative. It became known as the "God Committee" and is still in the law today.

On January 23, 1978, the Senator's God Committee ruled unanimously against the dam. Some members of Congress, with some adroit log-rolling, managed to insert an amendment into a pork-laden appropriations bill that authorized the TVA to finish the dam regardless of any other law. On November 29, 1979, the flood gates on the dam were lifted. A year later David Etnier, the man who started it all, found the snail darter in tributaries downstream from the dam. He recommended that the Fish and Wildlife Service remove the darter from the endangered list, which it did in July of 1984.

The battle over the Tellico Dam made it abundantly clear that any project—dam, highway, shopping center, parking lot, or timber sale—would have to take care of "every creeping thing that creepeth upon the earth." Far away from the Tennessee River in the forests of the Pacific Northwest another battle was brewing that made the Tellico Dam a small battle indeed.

As with the snail darter, a university student and a faculty member first raised concern about the Northern spotted owl. Eric Forsman, a graduate student at Oregon State University, began studying the biology of the spotted owl in the 1970s. His advisor was Howard Wight Jr., the son of the distinguished wildlife professor at the University of Michigan's School of Forestry during the 1930s and 1940s.

Forsman's data showed that all of the many pairs of owls he located were in old growth timber, and that in most cases the owl sites were on timber sales laid out for cutting. By March 1972, Forsman was convinced that the owl was in trouble unless Forest Service and Bureau of Land Management (BLM) practices were changed. Steven Yaffee of the University of Michigan, in his book *The Wisdom of the Owl,* quotes from the letter that Howard Wight sent to the Fish and Wildlife Service and to the region officials in the Forest Service and BLM:

> Forsman has located 37 sites where pairs of spotted owls can regularly be seen. All of these pairs are inhabiting old-growth timber. No other spotted owls had been located in any other habitat. These birds seem to require climax forests and are very sensitive to any alteration of these habitats. Preliminary observations indicate that if a forest habitat is altered the birds are forced to move, and do not reproduce the following year. They move to the nearest old-growth they can find, but with current timber management practices in Oregon this is becoming more and more difficult.

Wight's letter was duly noted by the agencies, but for awhile at least it was Forsman leading a lonely fight. Yaffee's book chronicles the progress of the controversy.

The fireworks began throughout the timber country of the Northwest when the owl was listed as "threatened" under the Endangered Species Act; sparks flew again when the Interagency Scientific Committee reported its findings that would result in a 30 to 40 percent reduction in timber sale volume from Forest Service and BLM lands; and then again when Judge William Dwyer in Seattle enjoined almost all timber sales from federal lands in Washington and Oregon.

The resulting gridlock brought angry demonstrations in mill towns throughout the region. People spent millions of dollars on lawyers, lobbyists, and advertising campaigns. It was jobs versus owls as charges and countercharges were hurled from the opposing camps. Secretary of the Interior Bruce Babbitt called it a "train wreck." But just as the snail darter was used in attempts to stop a dam, the owl was being used to stop the cutting of old growth forests on federal lands in the Northwest. These forests were now being called "ancient forests," and their management by Forest Service and BLM foresters was the subject of front page stories in newspapers and national network television shows across the land.

On May 6, 1991, the Fish and Wildlife Service listed the land they identified as critical habitat for the endangered owl. It totaled 11.6 million acres and included three million acres of private land. The report indicated that there might be significant impact on private lands. Although Weyerhaeuser at Coos Bay had never purchased any federal timber, it now became apparent that operations on the Millicoma Forest could be seriously threatened if any owls were found in the forest. And the odds were very good that owls would be found in the Millicoma.

Meanwhile the country was gripped in a presidential campaign. Bill Clinton, the Democratic candidate, had promised that he would break the gridlock in the Pacific Northwest. The industry looked to him to provide timber for their starving mills. The environmentalists looked to him to save the ancient forest ecosystem.

After his election, in an attempt to fulfill his promise, Clinton called for a Forest Summit, which was held on April 12, 1993. The summit was viewed as an important political event. One congressional aide stated, "This is going to make setting up the Middle East peace talks look like a Sunday picnic." It was truly a remarkable event. For a long day the president, the vice president, and five cabinet members listened to a long parade of experts and people affected by the controversy. One participant told the president, "I may be wrong, but I think this is the first time in Oregon history that the president, vice president, and five cabinet members have all visited the state at one time and in the same place, and we are honored." One of the participants in the conference was Charles W. Bingham, then executive vice president of the Weyerhaeuser Company and a native of Coos County. Bingham praised the leadership exhibited by the president in calling the conference.

Not since Teddy Roosevelt had a president of the United States taken such an active role in forest policy. At the conclusion of the conference, President Clinton charged his cabinet secretaries to come up with a balanced and comprehensive plan to end the stalemate, and he wanted it completed within sixty days. He said, "The process we have begun will not be easy. Its outcome cannot possibly make everyone happy. Perhaps it won't make anyone completely happy. But the worst thing we can do is nothing."

For the next sixty days, behind closed doors, some six hundred scientists, technicians, and support personnel from all the agencies involved, worked almost night and day. The group was named the Forest Ecosystem Management Assessment Team. FEMAT, as it became known, was chaired by Jack Ward Thomas, who a few years before had chaired the Interagency Scientific Committee, which had developed the report, "A Conservation Strategy for the Northern Spotted Owl." Later, Thomas was named Chief of the United States Forest Service.

The FEMAT report, which made the president's deadline, was titled, "Forest Ecosystem Management: An Ecological, Economic and Social Assessment." At almost one thousand pages and weighing several pounds, the report was probably the most exhaustive study ever made of a forest ecosystem. It developed ten options that had the objective of producing management options that complied with existing laws and met environmental, economic, and social needs of the region. Option 9 was the recommended option of the team. It set aside millions of acres of "Late-Successional Areas" and "Riparian Reserves" where no timber harvesting would occur. Other areas where some harvest could occur were Managed Late-Successional Areas, Adaptive Management Areas, and Matrix. The total annual cut from these sources would be 1.2 billion feet for the region. The average for the previous decade had been 4.6 billion feet. As the president predicted, no one was happy. In fact, it seemed that everyone was just plain angry. The industry dependent on federal timber was aghast. The environmentalists were suspicious of "adaptive management," which they thought was just another cover for the Forest Service to clearcut their ancient forests. The gridlock seemed as tight as ever.

The team when submitting its report said, "We have done our best to fulfill the charge given to us in the time allotted. . . . Our work as scientists, economists, analysts, and technicians is complete. Whatever decisions that may emerge from this work are now, most appropriately, in the hands of elected leaders." In July 1993, the president adopted Option 9 as the administration's plan. The controversy raged on, but the federal courts finally blessed the plan. Whatever the final outcome, management of the federal lands in the Northwest had been profoundly changed. What did this mean for our Millicoma Forest?

References

Forest Ecosystem Management Team. *Forest Ecosystem Management: An Ecological, Economic, and Social Assessment.* Washington, D.C.: FEMAT, 1993.

Interagency Scientific Committee. *A Conservation Strategy for the Northern Spotted Owl.* Portland, Ore.: Interagency Scientific Committee, 1990.

Mann, Charles C., and Mark L. Plummer. *Noah's Choice: The Future of Endangered Species.* New York: Alfred A. Knopf, 1995.

Sher, Victor M., and Andy Stahl. "Spotted Owls, Ancient Forests, Courts and Congress." *Northwest Environmental Journal* 6 (1990): 361–84.

Yaffee, Steven Lewis. *The Wisdom of the Spotted Owl: Policy Lessons for a New Century.* Washington, D.C.: Island Press, 1994.

"We're trying to protect the right to grow and harvest wood from some of the world's most productive forestland. To do that we have to protect the soil. We have to provide wildlife habitat. We have to understand our relationship with public landowners for protecting threatened species. It's a new mindset and may be more important than any single change in industrial forest management in the last 94 years."

—Charles W. Bingham, 1995.

17 / A New Look

One of the goals of the Weyerhaeuser Company was to be in the top quartile in profitability of the relevant competitors in the industry. Beginning in 1988, the company examined each of its businesses rigorously. Timberlands came under the microscope as well. There was no doubt that this business was the leader in returning net cash to the corporation. It also was a leader in having the lowest costs incurred in establishing and managing the new forest, and it certainly excelled in its use of research. Where it was failing badly was its relations with the public. The company, through town hall meetings in metropolitan areas of the Northwest and many focus groups, found that it was out of sync with much of its public.

What Weyerhaeuser found shocked its senior management, and especially its foresters. The vast majority of the company's public was skeptical of forest practices on industry lands. It was concerned about the beauty of the landscape, biodiversity, and clean streams that supported fish. And it did not like the rate of clearcutting of ancient forests. Much of the public thought tree farms were sterile monocultures. What the industry considered commodities most of the public considered "precious resources." The practices of "Weyerhaeuser—The Tree Growing Company" that made it a hero in the 1960s and 1970s no longer represented the values of a changed society. More alarming to management were findings that the public did not differentiate between management of privately owned industrial forestlands and that of the public lands managed by agencies such as the Forest Service and BLM.

Probably no one in the Weyerhaeuser hierarchy was more concerned than Charles W. Bingham, then the executive vice president of the corporation. Throughout his career with the company, he had been an agent for change and was about to direct another fundamental change in the corporate culture.

Charley was reared in Myrtle Point, Coos County, in a family dependent on the logging economy. As a boy, Charley—blond, blue-eyed, with a dimpled smile—may well have been the smartest kid in town. He was class valedictorian at Myrtle Point High School. Not many, if any, Myrtle Point high school graduates went to Harvard, but Charley did. He graduated magna cum laude from both Harvard and Harvard Law School. He joined Weyerhaeuser right out of Harvard, and when he retired in 1995 after over thirty years with the company, he was magna cum laude again.

Beginning in the law department, Bingham rose through one management job after another. He became a legend within the company, and next to the CEO he was, for much of his career, the most powerful man in the company. Intellectually arrogant, having little patience with subordinates who did not share his vision, he drove himself and his staff unmercifully. A comment from Charley could bring strong managers to tears.

Through good times and bad, Charley always supported the forestry budgets. Now he was concerned that with all the huge investment in high yield forestry the company would lose the ability to reap the benefits. It was time for a new look. Asking the question, "What must be done for Weyerhaeuser Company to be successful in private forest management in the 21st Century?," he formed a task force and appointed Peter Farnum, the company's director of Environment and Forestry, to head it up. The task force created a stewardship vision that committed the company, in addition to producing wood, to protect, maintain, or enhance values such as water quality and fish habitat, wildlife habitat, including threatened or endangered species, soil productivity, aesthetics, plant and animal species diversity, and cultural or historic areas.

This multidisciplined task force also found that the public—both locally and at large—must be listened to. Local forest councils were created to test alternative harvest methods, species composition, and management for the resources found in the forest other than trees. The only "given" to the task force was that the company would continue to practice even-aged management, which meant clearcutting, even though the public found this practice objectionable. However, Forestry 21 was going to carefully examine the size, shape, and location of harvest areas, in some cases with public input.

Bingham, along with Pete Farnum, Donna Brown, and the regional vice presidents, held meetings across the country with every operating forester, manager, and research scientists. Managers were evaluated on their personal support of what Charley Bingham called an "important paradigm change." Those within the company were observing changes in Charley as well. Some called it the "greening of Charley Bingham." But he never lost sight of the fact that whatever they did had to be economically sound.

The program was recognized that the program was a long-term commitment. Any change claimed had to occur in the forest, and it had to be sustained and credible. An early visible success story was the cooperative watershed planning project on a major watershed in the state of Washington, which was highly praised by public agencies.

While Forestry 21 was long term, the short-term political pressures on private lands in the Northwest were hanging over the forest like a dark and very ominous cloud. Weyerhaeuser's senior management and most notably, Charley Bingham, were concerned that the risks to the ability to grow and harvest trees and make a profit in a free, global market were growing greater every day.

The Northern spotted owl was the cause of most of the clouds. When the owl was listed as threatened under the Endangered Species Act, most people thought the action was aimed at public lands in the Northwest and the cutting of old growth on the national forests and BLM lands. It soon became apparent, however, that the protocol for protecting the owl did not stop at ownership lines. With 2.2 million acres of forestland in the Pacific Northwest, Weyerhaeuser was more at risk than any other private landowner. When Eric Forsman first studied the owl in 1972, he located thirty-five pairs in old growth forests. By the time of the Interagency Scientific Study in 1990, the population of owls was estimated at between 3,000 and 6,000, but no one was really aware of this.

By now countless biologists and technicians were looking for owls on federal and private lands. Owls seemed to be showing up everywhere. In the dark of night, crews would head out into the forest calling for them. These tame little birds would readily come to the caller who would feed it a mouse. On at least one site on the Millicoma, the owl would appear at the sound of a slamming pickup door.

Weyerhaeuser biologists were learning a lot about the spotted owl, but the law had to be obeyed. Owl circles had been established by not only the federal regulations, but also by the states of Washington and Oregon. These circles, which surrounded known owl sites, could involve thousands of acres in which no activity could take place. To their horror, Weyerhaeuser managers found that some 400,000 acres of their productive forest land were off limits to any harvest. With thirty-five owl pairs and individual owls residing on the Millicoma Forest, the future of the new, efficient CBX mill, as well as other mills that depended on the Millicoma for portions of its log supply, was in jeopardy. Something had to be done.

References

Bingham, C. W. Correspondence with the author, 1996.

———. "Evolution in the Forest." *Weyerhaeuser Today,* March 1995.

Clarke, James. Interview with author, Coos Bay, Oregon, 1995.

Interagency Scientific Committee. *A Conservation Strategy for the Northern Spotted Owl.* Portland, Ore.: Interagency Scientific Committee, 1990.

Weyerhaeuser Company. *Project Legacy.* Presentation to the Senior Management Team. Tacoma, Washington, 1993.

"Weyerhaeuser has distinguished itself as a real leader in the future of forest management. The land included in this plan will now be managed for both timber and owls, showing that we can achieve our conservation goals and still cut timber in an environmentally responsible way." —Secretary of Interior Bruce Babbitt, February 14, 1995.

18 / Habitat Conservation Area

While Forestry 21 had captured the imagination and enthusiasm of Weyerhaeuser foresters, it was becoming increasingly clear that public values were still not in tune with Weyerhaeuser's goals. It was time for another task force. Named "Project Legacy," the new team was made up of people from timberlands, law, government affairs, communications, and innovative thinkers from anywhere within the company. Their overall goal was to regain the initiative in public policy by clearly differentiating private industrial forestland management from public forestland management in the Douglas-fir region of Washington and Oregon.

For years Weyerhaeuser had used the slogan, "The Tree Growing Company," which indeed it was. The task force research showed, however, that tree planting alone did not meet the concerns of much of the public. People still felt the company was running out of wood, that industry encouraged overconsumption and pursued profit at the expense of the environment and the local community. The task force pointed out that the company needed to demonstrate it was listening and changing, and that it supported and acted on a shared conservation ethic with the public.

The company began taking a holistic view of its forests. It also stepped up its recycling program. Soon the company's advertising program stressed "Weyerhaeuser, Your Partner in Recycling." As consumers in metropolitan centers across the country were going to the curb with their piles of old newspapers, much of it was winding up in Weyerhaeuser recycling plants. The company had become probably the largest recycler of paper products in the country.

Meanwhile the situation on the Millicoma was becoming critical. It was ironic that even though the closing of the big mill on the Coos Bay waterfront in 1989 had little to do with the spotted owl, now that the transition from old growth had been pretty much successfully bridged, the owl threatened the new mill.

The alternatives did not look very promising. Many were advocating litigation. This was deemed expensive, much too long a process, and with no guarantee of success. Another alternative was to get the Endangered Species Act amended. Even though evidence proved that Congress had not intended to cover management practices that might indirectly affect a species or impact private lands in such a major way, Congress had demonstrated that they did not want to handle this hot potato. Even if they did, there was no guarantee that an administration would support any change.

Project Legacy had indicated that the company should listen to the public. All the evidence indicated that the public did support protecting endangered species. Perhaps the company had best try to find a way to raise owls as well as trees. Was there a way, within the law, that this could be done?

The Endangered Species Act was amended in 1982 by adding a Section 10, which allowed the Secretary of the Interior, through the Fish and Wildlife Service, to issue a permit for the taking of a threatened species if the taking is incidental to an otherwise legal activity, and if the applicant prepares a Habitat Conservation Plan (HCP) that minimizes and mitigates the impact of the taking to the maximum extent practicable. The plan must not appreciably reduce the likelihood of survival and recovery of any species in the wild. The decision was made to apply for a HCP permit, knowing the cost would be high and the trail torturous.

There seemed to be no other way out of the crisis. Bingham had been disappointed that President Clinton at the Forest Summit meeting in April of 1993 seemed unaware of the regulatory impact on private lands. The following month, Bingham and company CEO Jack Creighton traveled to Washington for a meeting with Babbitt and Secretary of Agriculture Mike Espy. When told of Weyerhaeuser's commitment to pursue an HCP, Babbitt was encouraging. Espy, according to Bingham, "never showed any interest, period."

By that fall many meetings had been held with top staff of the Department of Interior and the Fish and Wildlife Service. Secretary Babbitt had realized that it was as much in his interest as Weyerhaeuser's to see the HCP process work. The Endangered Species Act was under heavy fire from the Republican Congress, and the administration needed something to show that they were resolving what Babbitt called "the complicated natural resource problems which had plagued the Pacific Northwest for so long."

Two days before Christmas of 1993, Charley Bingham traveled alone to Portland to meet with the Regional Director of the Fish and Wildlife Service, Michael Spear. This was the beginning of the formal negotiations. Charley told Spear that the company wanted to be a part of the solution, and they were dedicated to making the Endangered Species Act work. He pointed out that Weyerhaeuser's sustained cut in the Pacific Northwest was greater than President Clinton's Option 9 for the entire western National Forest System. He also demonstrated the acute situation at Coos Bay. He found a sympathetic audience in Spear.

After the decision was made to go ahead with the Habitat Conservation Plan, Weyerhaeuser put together a team led by forester Bruce Beckett in Tacoma.

Curt Smitch of the Fish and Wildlife office in Olympia, Washington, headed the HCP, and Robin Bown of Fish and Wildlife's Portland office was the Oregon contact.

Much of Weyerhaeuser's logging on the Millicoma was now in the Vaughan area, some of which had been logged around the turn of the century. These stands had not been involved in the eighteenth-century fires, and logging had come back slowly over the years. Most were from sixty to eighty years old but contained a few larger trees. Although these stands were not old growth, they were made up of a layered structure that made for ideal owl habitat. Not only did the owls require breeding habitat, but they also needed foraging areas and dispersal habitat for the young birds to establish new home ranges. Connectivity between the known owl ranges throughout the region was a vital concern for the bird's survival, according to the federal biologists. The Millicoma Forest was ideally situated to provide the link, since the forest was bounded on the west by the Elliott State Forest and on the east and south by federal forestlands, both of which contained "Late Successional Reserves." These reserved areas of old growth forests contained forty-four pairs or single owls within 1.5 miles of the Millicoma's boundaries.

Weyerhaeuser agreed to reserve for the next twenty years at least 1,963 acres of mature timber valued at $40 million. They also agreed to manage for dispersal habitat, which would limit large gaps between owl habitat areas. Almost all of the gaps would not exceed three miles. Known owl sites outside the reserved area would be closely monitored, and no harvesting or road building would occur within a quarter of a mile of an active nest during breeding season. The agreement would remain in effect for fifty years and could be renewed for another thirty years after that.

Before any of this could be approved, an environmental assessment under yet another federal law, the National Environmental Policy Act (NEPA), had to be provided to the Fish and Wildlife Service. On November 16, 1994, the "Environmental Assessment for the Proposed Issuance of a Permit for Incidental Take of the Northern Spotted Owl" was delivered to the U.S. Department of Interior, Fish and Wildlife Service in Portland, Oregon. It was prepared by Beak Consultants of Kirkland, Washington, and was paid for by the Weyerhaeuser Company. The company considered the requirement for an environmental assessment as the most onerous part of the entire process. The assessment covered a good deal more than the owl. It looked at everything from the Millicoma longnose dace to tall bugbane. It examined various alternatives, and the proposed habitat conservation plan proved the strongest.

Finally, on August 13, 1995, the agreement between Weyerhaeuser and the U.S. Fish and Wildlife Service was announced. The plans for Secretary Babbitt to come to Portland for a public ceremony had to be canceled due to a blizzard that shut down everything. The government's press release quoted Babbitt: "This agreement is the first of its kind in Oregon, and it was successful because of Weyerhaeuser's strong commitment to making it work." At a very high cost Weyerhaeuser had brought some certainty to the continued management of the Millicoma Forest. It had responded to much of the public's demand that a forest is more than trees.

The company was still at risk, however. In a statement to the U.S. Senate Subcommittee on Environment and Public Works, on August 3, 1995, Jack Larsen, Weyerhaeuser's vice president for the environment, pointed out some of the risks and lack of certainty that make the process less than user friendly. Chief among these was the species-by-species habitat conservation plans instead of a habitat-based approach that covered multiple species. No landowner could operate on the threat of an "endangered species of the week."

Larsen was critical of the need for an Environmental Impact Statement (EIS) under NEPA. Larsen called it "a serious impediment to landowner participation in the HCP process." He added that "this requirement adds significant costs, time and risks to applicants. In fact, more than half of the cost and time to develop Weyerhaeuser's Coos Bay HCP were directly related to EIS requirements under NEPA. In addition, the NEPA documentation cannot be completed until the HCP is nearly approved, but the HCP cannot be finalized until the NEPA documentation and process have been completed. Therefore applicants are faced with uncertain outcomes, uncertain timelines for completion, uncertain costs, and a risk of legal challenge to the NEPA documentation. These uncertainties create serious disincentives for landowners to develop HCPs." Larsen asked the committee to consider exempting the requirement for an EIS in the HCP process.

It was now forty-five years ago that the first tree was felled on the Millicoma. None of the loggers or foresters working then had any concern about the Northern spotted owl. Today the foresters and loggers are raising owls along with trees. As Charley Bingham had said, there was a new mindset in the land.

References

Beak Consultants. *Environmental Assessment for the Proposed Issuance of a Permit for the Incidental Take of the Northern Spotted Owl.* Millicoma Tree Farm, Weyerhaeuser Company, Coos and Douglas Counties, Oregon. Portland, Ore.: U.S. Fish and Wildlife Service, 1994.

Beckett, Bruce K. "Improving the Habitat Conservation Planning Process." Habitat Conservation Planning Workshop, August 10, 1995.

Bingham, Charles W. Correspondence with the author, 1996.

Larsen, Jack. 1995. Statement for the Record. U.S. Senate Environment and Public Works Subcommittee on Drinking Water, Fisheries and Wildlife. August 3, 1995.

U.S. Fish and Wildlife Service. News Release. "Another Government-Industry Agreement on Owls: Interior Secretary Babbitt and Weyerhaeuser Launch Conservation Plan—A First for Oregon." Portland, Ore.: Fish and Wildlife Service, 1995.

Weyerhaeuser Company. *A Habitat Conservation Plan For the Northern Spotted Owl. Plan Summary.* Springfield, Ore.: Weyerhaeuser Company, 1995.

"It seems reasonably certain that whatever might be the fate of the forests which now cover western Washington territory and Oregon they will be succeeded by forests of similar composition and that this whole region ill adapted in soil and topography to agriculture will retain a permanent forest cover long after other great forests of the continent have disappeared." —Charles Sprague Sargent, 1890.

19 / The Millicoma Forest: Today and Tomorrow

The Millicoma Forest, born by the forces of nature and tended by the human hand, now flourishes. The clearcuts of the 1950s and 1960s are now an unbroken expanse of green. The roads that once shook under the weight of huge logging trucks are now tunnels through a solid wall of young trees. Many parts of the forest are silent, rarely visited by anyone. No public highways cut across the forest, but there are now over two thousand miles of company roads. There are no Wal-Marts or McDonalds. This is tree-growing country—it always has been and probably always will be. The area at the top of the 1040 line that was planted in 1953 now has trees reaching 120 feet tall and a foot and a half in diameter. The landings where a person could look out for miles across the patchwork logging are now completely closed in. Over 170,000 acres of the forest are classified as "Early-Successional" up to forty years old. Another 25,000 acres are in "Mid-Successional" from forty to seventy-nine years old. Mature and old growth forests make up only 11,000 acres.

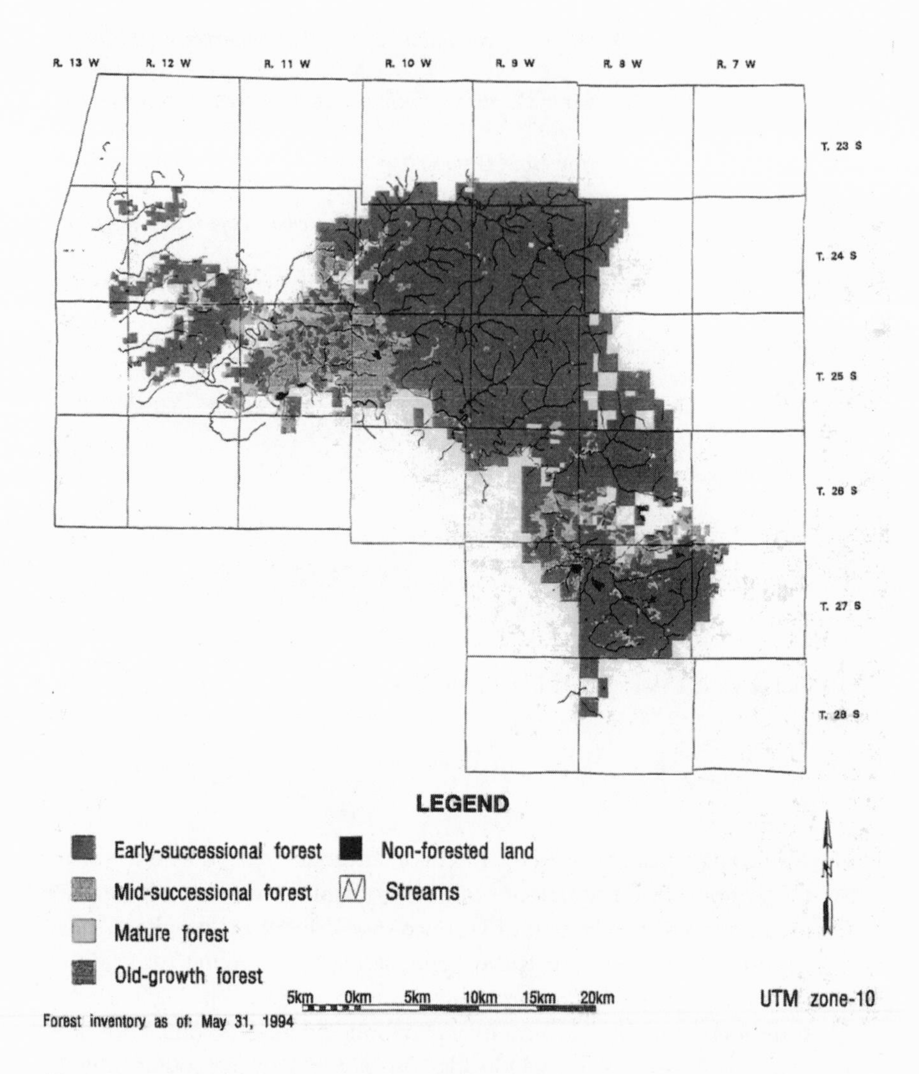

Forest cover types on the Millicoma Tree Farm in 1994. Source: Beak Consultants Environmental Assesments.

All of the forest is managed under the strict rules of the Oregon Forest Practices Act, which include extensive riparian management rules. Clearcut size is limited and adjacent areas cannot be logged until trees on the logged area reach at least five feet in height. Owl populations are monitored by federal and company biologists as are streams. Not only does the forest support the spotted owls, but it also has bald eagles, pileated woodpeckers, egrets, great blue herons, and a variety of song birds. Plants range from lichens and liverworts to ferns and species that conventional wisdom had assigned only to old growth forests.

The forest is laced with countless streams, rivulets, and several rivers. Approximately 114 miles of fish-bearing streams are located in the Millicoma Forest. These streams support runs of fall chinook salmon, coho salmon, chum salmon, winter steelhead, and sea-run cutthroat trout. Weyerhaeuser scientists and state fisheries biologists have conducted several fish habitat and stream surveys in the forest including Weyerhaeuser's watershed analysis of the East Fork of the Millicoma. In 1961, the Oregon Fish and Wildlife Department, along with Weyerhaeuser, had built a salmon-raising pond in an oxbow of the Millicoma River created when the company replaced two bridges on the blacktop with culverts. After several years, because of disease and predation, the state abandoned the Millicoma Pond, as it was called. Some thirty years after everyone had forgotten about it, the state and Weyerhaeuser built a concrete fish ladder that deepened and enlarged the pond, which over the years had shrunk considerably. Also three boulder weirs were built in the East Fork of the Millicoma near the junction with Matson Creek. The weirs were designed to create gravel-spawning beds and create pools to slow down the winter freshets. Today large chinook salmon are spawning in the Millicoma.

And what about the Eastern brook in Matson Creek? Sadly, it seems that the actions of an unknown game commission biologist may have inadvertently wiped out these unique fisheries. In 1962, Robert Corthell, with a backpack can full of cutthroat trout, hiked above the falls of Fall Creek and planted the trout. Like Matson Creek, Fall Creek had no resident fish because of the falls. And like the Eastern brook in Matson Creek the cutthroat trout planted in Fall Creek flourished. Corthell, then the biologist for the Fish and Game Department in the Coos Bay district, was well aware of the Eastern brook in Matson Creek. He had first heard of the success of the species in Matson Creek from Professor Dimmick at Oregon State College in 1947. Corthell (along with his knowledge of the Eastern brook) was transferred to Portland. He feels that someone impressed with the success of the Fall Creek planting introduced cutthroat into Matson Creek. According to Corthell the cutthroat outcompetes other fish when occupying the same stream. Therefore, the Eastern brook lost out.

Elk are still found throughout the forest, even in some of the densest young stands. While it was common in the 1950s to see large herds in the clearcuts, now the elk are seen in groups of three or four. Beaver are at work in Matson and Conklin Creeks, and cougar sightings are on the increase.

On a board foot basis, the forest is now producing about 100 hundred million board feet a year from the far reaches of the Callahan area and the sixty- to seventy-year-old stands in the Vaughan area. Sometime within the next ten years the cut will rise to 200 to 250 million feet, which can be sustained at that level. The young stands along the 3000 line are now being thinned. In order to minimize damage to the remaining stands, the loggers must be very careful and consequently only produce two loads of thinnings a day.

At Allegany, the big old building that housed the store and post office is gone. The Allegany School is no longer used as a school but serves as a community center and is also the site of the Allegany Post Office, which on March 25, 1993, celebrated its centennial with a special postmark. There is also a new convenience store in Allegany that features a wall devoted to Polaroid pictures of the proud hunters and fisherman posing with their elk, deer, or steelhead.

The road turnout, which overlooked the Allegany log dump, was a favorite spot for tourists to watch the logs splash into the river. It featured an attractive sign that told the story of the never-ending stream of logs coming down from the Millicoma Tree Farm. One of the crew burning down the terminal in 1989 remarked, "We better burn this too," and it went into the flames. Also at the turnout was the bronze plaque commemorating Schenck's dedication of the tree farm. It had been missing for sometime until someone found it on the roof of the schoolhouse.

Downtown the crummies still leave for the woods. The attractive Weyerhaeuser office still looks out at the huge building that once housed the most modern plywood plant in the country. But instead of the hissing of the giant presses, now one hears the cha-ching of slot machines and the click of gambling chips.

After Weyerhaeuser sold the plywood plant to Fred Sohn, he too found that it was a beautiful but unprofitable plant. He sold the plant to the Coquille Indians, who claimed the land was theirs anyway. They planned to convert the plant into a gambling casino. According to a story in the *Wall Street Journal* dated February 14, 1995, "The casino, as big as two football fields, is to be called the 'Mill.' Giant saws, weathered old beams and other woodsy paraphernalia would remind gamblers that they are indeed in timber country." One of the tribe's partners in this venture was Lee Iaccoa, at one time one of the nation's most celebrated businessmen. The plan was greeted with controversy not only from the community but by another tribe in the region, the Coos Indians. The Coos claimed this was their territory and appealed to Secretary Babbitt to stop the Coquilles. He refused to enter the controversy, and The Mill is now in business. An ad for the casino shows a grandmotherly type exulting in front of a slot machine with the headline, "It Beats Knitting!" It must beat making plywood.

Up in the hills, Weyerhaeuser was gambling too. It was making bets far into the future that it could make a profit from a commodity in a free market economy when the same lands that produced the commodities also produced owls, woodpeckers, and scenery, which were difficult or impossible to price in such an economy. Meanwhile, on the national scene the gridlock over the old growth forests on public lands continued. The Sierra Club, one of the oldest conservation organizations in the country, called for a ban on any cutting of ancient forests on public lands. The distrust and outright hostility among interest groups prevented any rational discussion of the nation's forest policies. The only place the opposing sides met were in courtrooms.

Executive officers of the Weyerhaeuser Company were being asked, "In this environment what is the future of private forest investment?" George Weyerhaeuser stated that "timber growing is obviously a business for patient people and patient capital." Would the patience pay off? Following President Clinton's Forest Summit meeting in Portland, two of the participants met for dinner—Charley Bingham and John Gordon. Gordon is a distinguished forest scientist, a Yale professor, and former dean of the university's School of Forestry. For some years he had been trying to inject science and research into the increasingly rancorous debate. In 1990, he headed the study of the National Research Council that produced the report, *Forestry Research: A Mandate for Change*. The report called for research with an environmental paradigm. In 1991, two congressional subcommittees asked for the convening of a panel of scientists, which Gordon headed, to come up with some options for solving the owl controversy and the economic cost of each of the options. Known as "the gang of four," their report came up with fourteen options starting from the assumption that nothing less than the findings of the Interagency Study would ensure the survival of the owl. The congressmen found nothing in any of the options to get them off the hook, and the interest groups on both sides howled bloody murder. The later FEMAT study used some of the strategies developed by the gang of four.

These two men meeting over dinner in Portland were the catalysts for what became the most important meeting on forests since Teddy Roosevelt's Second American Forest Congress in 1905. They explored these questions: "How can we determine what the American people want from their forests, both public and private? Is there any way to dispel the distrust and hostility between the interest groups? Can we find areas of agreement rather than continue to fight over our differences?" The decision was made to try to get a diverse group of people together and explore a process to get some answers.

Charley Bingham assigned the task to Rex McCullough, the vice president of forestry research who headed the largest private silvicultural and environmental forestry research staff in the world. Bingham secured the blessing and full participation of Jack Creighton, his CEO. Gordon rounded up a cadre of his students and received the backing of his dean, Jared Cohon. Soon an ever-expanding group of men and women were meeting regularly at sites across the country under the auspices of what became known as the Yale Forest Forum. Gordon and McCullough served as co-chairs. Enthusiasm for action was growing, although it was still not clear exactly what kind of action they would take.

In January 1995, the Yale Forest Forum convened a Forest Roundtable at Nebraska City, Nebraska. This small town in the middle of the winter seemed a most unlikely spot for a meeting on forests. The participants had been carefully selected to represent as diverse a group of people as possible all with a stake in America's forests. It did not include the "usual suspects" i.e. industry associations. The decision had been made to hold a major conference called the Seventh American Forest Congress, building more on the historical context of the name rather than on the structure. The previous congresses had been held at critical times in our forest history beginning in 1882 and the last in 1976. The Nebraska participants were asked to define the key components of national forest policy that reflects a shared vision; identify the problems and issues that are impeding a shared vision, develop criteria for evaluating solutions, and finally, to structure a forest congress to develop visions and principles. The conveners of the roundtable held their breath, for if this group didn't agree that a forest congress should be held, the entire effort was dead in the water. Not only did they enthusiastically endorse the plan, but they also strongly recommended that local roundtables be held in communities across the country preceding the congress.

It was finally agreed that the congress would be held in Washington, D.C. in February 1996. Now it was put up or shut up—and indeed a short time for such a huge undertaking. An executive director, William Bentley, was selected, and an office at Yale was established with Robert Clausi as manager. Everyone else involved were volunteers with regular day jobs. The board of directors now had members as diverse as the Wilderness Society, Audubon Society, community-based environmental groups, state foresters, industry representatives, and academics. They raised money from foundations, industry, federal agencies, and individuals. A sizable percentage of the money was used for scholarships for people who could not afford a trip to Washington, principally community-based environmental activists.

By the time of the congress, over fifty roundtables had been held throughout the country. Each had developed vision statements, and most had agreed on a set of principles. The design team for the congress had proposed a bold plan that was unlike any of the former congresses. First, this was to be a citizens' conference, open to all. Next, there would be no speeches or papers or preconceived outcome. And finally, the participants would sit together at round tables of no more than ten people, with a chart board at each table. The motto of the congress was "From Many Voices—A Common Vision." For three and a half days each table with their own facilitator would hammer out a set of visions for the forests of the twenty-first century, a set of principles to implement the vision, and next steps.

For several days before the opening of the congress, registrants were streaming into the huge Washington Sheraton Hotel in the Woodley Park section of the nation's capital. Some had ponytails and carried backpacks, and some had blue suits and carried briefcases. They came from every state except Kansas and Rhode Island. They ranged from octogenarians to high school students. There were people concerned about the mean streets in cities and people concerned about logging towns in the High Sierras. Participants ranged from landowners with forty acres in Mississippi to corporation executives responsible for millions of acres across the land. The Chief of the U.S. Forest Service and agency people from the Department of Interior also attended. Some people may have spiked trees or chained themselves to bulldozers. This truly remarkably diverse group of people had one thing in common—a deep concern about the forests of America.

On the morning of February 20, fifteen hundred people sat among the sea of tables that covered almost every foot of the grand ballroom of this huge hotel. This had all been put together by volunteers. If de Tocqueville himself had looked in at this scene he might have exclaimed, "Mon Dieu! Only in America." In fact there were observers from several foreign countries who came to see what "these crazy Americans are up to." After some initial posturing and feeling each other out, the participants, seated in as diverse a mix as possible, began to work on the process.

The forest historian Char Miller described his table this way: "This representative mix might have confounded our capacity to devise vision statements, develop principles by which these visions would be framed, and delineate some practical first steps to achieve these goals, but this did not occur because our intense work was conducted almost exclusively at the tables. That context forced us both to talk and listen, to argue and negotiate, and to do so all within the confines of a round table. . . . It's hard to turn your back when you are looking into their eyes." Brock Evans, the veteran environmental lobbyist for the Sierra Club and Audubon Society, said, "At least we are not meeting in a courtroom." By the end of the congress, most people who shared a table had bonded. They hugged and exchanged warm handshakes and addresses. If nothing else, a dialog had begun.

The congress participants agreed that in the future our forests would be held by a variety of private and public ownership whose owners' rights, objectives, and expectations are respected and who understand and accept their responsibilities as stewards. One of the principles adopted included the following language: "All forest owners acknowledge that public interests (e.g. air, water, fish and wildlife) exist on private lands and private interests exist (e.g. timber sales, recreation) on public lands." Sustainability, biological diversity, and productivity were all supported for the forests of the future. Was the gridlock broken on forest policy? Probably not. Will forestry for the twenty-first century be different from that of the twentieth? Most certainly. Charley Bingham said, "I feel very good about where we are. I only wish I were twenty years younger so I could participate more in the most exciting era of American forestry ever." Steven Yaffee said, "While we can bemoan the failings of our agencies and decision making processes as revealed in the spotted owl case, the response to the controversy can be a springboard to this exciting future." And Jim Clarke, forester for the Millicoma, said, as he looked out at the green wave of young trees covering the hills, "These are exciting times."

It is now the year 2000, and we have followed the life of our forest for well over two centuries. The forest has changed as has America and the world. The Millicoma has been molded by all the forces of nature, by corporate strategies, and the laws of governments. But the trees go on, marching up the hillsides, reaching for the sky. Elk and deer still roam the forest. The wild cry of coyotes running the ridges can still be heard. Chinook salmon and steelhead trout run up the Millicoma every fall, and eagles soar high above the treetops. There is a mystery about the deep forest. The winds come across the bay and sweep over the ridge tops creating a sighing in the tops of the firs. It seems to carry voices from the past—explorers and trailblazers, surveyors and cruisers, foresters and loggers. Yes, a forest even talks, if you listen.

References

Bentley William R. Final Report. Seventh American Forest Congress. New Haven, Conn., 1996.

Bingham, C. W. Personal correspondence with author, 1996.

Carlton, Jim. "Behind the Wheel." *Wall Street Journal*. February 14, 1995.

Clarke, James. Interview with author, 1995.

Corthell, Robert A. Personal correspondence with author, 1996.

Williams, Michael. *Americans and Their Forests*. New York: Cambridge University Press, 1989.

Yaffee, Steven Lewis. *The Wisdom of the Owl*. Washington, D.C.: Island Press, 1994.

Epilogue

Decisions made in a global economy may have profound effects in areas half a world away. In today's world no community is immune to social and economic change. Such was the case in the Coos Bay area on January 13, 1998, when Weyerhaeuser announced the permanent closing of the export sawmill.

The closing came as no surprise to the two hundred workers in the mill; it had been closed since the previous November due to lack of orders. Nevertheless, many were hoping for a miracle. The mill, which opened in 1989, was designed to saw metric lumber for Japan's housing industry. By 1998, Japan was gripped in a depression, and housing construction plummeted. In addition, Japan's home builders were using more dimension lumber, laminated beams, and prefabricated beams, which eroded the demand for the post and beam lumber produced by the Coos Bay mill. At the same time the company closed the export lumber dock, the export chip facility, and sold the office to the Coquille tribe. Weyerhaeuser also laid off most of the loggers at Dellwood. The harvest from the Millicoma Forest would now be by timber sales to independent contractors.

After these strategic decisions, Weyerhaeuser's presence in Coos Bay was down to 160 employees at the container board mill and forty timberland employees at Dellwood—a far cry from the over one thousand mill and woods employees in the 1960s. John Anderson of the State Employment Department said, "The closure would have a significant impact on a community that averages a 9 percent jobless rate." Like almost all resource-dependent communities, Coos Bay has seen booms and busts. A century ago most of these timber-dependent communities were left with nothing but barren hillsides. However, in the hills behind Coos Bay lies the Millicoma Forest, one of the most intensively managed forests in the world. The growth in this new forest is prodigious, and as it matures it will be providing new economic opportunities.

Today the harvest from the Millicoma is approximately 85 million board feet in timber sales from the eastern reaches of the forest and the Vaughan and Lockhart areas. A majority of this timber is going east to Weyerhaeuser's mill at Cottage Grove.

Back in 1944, Bob Conklin and Minot Davis pondered the question of how to best develop this challenging forest property. The question then was, is it better to go up the South Fork and tap the old growth on the east side or up the Millicoma to harvest the red fir? The decision was to go up the Millicoma. Today most of the timber from the new forest is going east. Markets and economics will govern future flows from the Millicoma.

The young trees now covering the hills of the Millicoma Forest are oblivious to any boardroom strategies, and each year will add another ring and reach closer to the sky.

Glossary

Brow log: A large log at the top of the unloading ramp.

Board feet: A measure of the amount of wood in lumber, log or tree. A piece of wood one-inch thick and twelve inches square.

Booms: a line of connected floating logs or timbers enclosing logs stored in the water.

Booming rights: a lease or purchase of the right to establish booms along a property owner's waterfront.

Bull buck: foreman of the crews who fall the trees and cut them into logs.

Cat: A generic term for tractor derived from Caterpillar a major manufacturer of tractors. The size of the tractor is indicated by designations such as D-7, D-8, or D-9; D-7 being the smaller.

Cruising: A sampling method to determine the volume and quality of timber on a specific piece of land. Generally a two person crew with the compassman in the lead running a course through the woods with the cruiser following, estimating, or measuring the diameters and heights of the trees on regularly spaced sample plots or strips.

Increment borer: An instrument to remove a small, pencil-sized core of wood from a tree that shows the annual growth rings. It does not harm the tree.

Landing: A level area where logs that have been removed from the cutting area are loaded on trucks.

Peckerpoles: Small second growth logs or trees.

Rigging: The cables and lines used in high-lead logging. Crews that worked in the rigging included all the workers involved in getting the felled and bucked timber from the setting to the truck.

Setting: An area laid out for logging.

Site: A measure of the quality of the forest soil. In Douglas-fir it is determined by the height of the tallest trees in a stand at one hundred years. Site 1 is the highest quality. Site 5 the lowest.

Side Rod: The foreman or boss of a logging area.

Stand: A group of trees in a continuous area.

Spar tree: A tall, sound tree that, after being topped, has a large block hung from the top that carried the wire cables used to hoist logs up or down the hill to a landing. This type of logging was called high-lead because the logs were partially suspended off the ground as they were brought into the landing. Spar trees have now been superseded by mobile steel towers.

Yarder: A machine with drums holding the cables that when reeled in "yarded" the logs to where they were loaded on to trucks or rail cars.

Index

Allegany, town of, 39, 107–8, 110, 132
Allison, Rex, 30, 71–72, 75, 93–94

Babbitt, Bruce, 125, 126, 133
Baker, Gil, 84–85, 86
Bears, 101–2, 103
Beetle (Douglas-fir bark beetle), 22, 33, 59–64
Bingham, Charles: as executive vice president of Weyerhaeuser, 121–22; meeting with Cabinet members, 125–26; as participant in 1993 forest summit, 118; quote from, 128; setting up Yale Forest Forum, 133–34

Calapooya Indians, 4
Champoeg Compact, 8
Chemical spraying, 90–92
Clemons Tree Farm, 44, 98
Clinton, Bill, 117–18, 125, 133
Coleman, Ovie, 42, 48, *49,* 54, 56, 103
Conklin, Robert P.: on dedication of Millicoma, 43; early career at Weyerhaeuser, 28–29, 34; hiring Herm Sommer, 40; leaving Coos Bay, 51; on logging, 71, 110, 140; and road construction, 33, 36; working with Minot Davis, 30–32
Coos Bay, description of, 57
Coos Bay Company, 17
Coos Bay Growth and Yield Study, 23–25, 68
Coos Bay Indians, 6, 133
Coquille Indians, 133, 139
Cornelius, Royce, *46:* collapse of his hose tower, 110; early career of, 34; hiring of, 33; influence of in Allegany, 39–*41;* opinion of tree farm, 43; transfer to Tacoma, 57

Cougars, 101–2, 103

Davis, Minot, 29–*30*, 31, 140
Donation Land Act, 10, 17
Douglas, David, 13, 16

Elk (Roosevelt elk), 65–67, 68, 100–101, 132
Endangered Species Act, 114, 125
England. *See* Great Britain
Espy, Mike, 125

Farnham, Thomas Jefferson, 8
Fire, 2–5, 77. *See also* Ivers Peak Fire
Fish: cutthroat trout, 102, 103, 131–32; Eastern brook trout, 103–5, 131–32; salmon, 102; snail darter, 115–16; steelhead trout, *102*, 103
Floods, 77
Forest Ecosystem Management Assessment Team (FEMAT), 118–19
Forest Lieu Act, 10
Forestry 21, 121–22, 124
Forsman, Eric, 116–17

Gaines, Bill, *45, 46,* 49–50, 55, 68–69
Gambling, 132–33
Golden Falls, *21*
Gordon, John, 133
Great Britain, negotiations with U.S., 7–8

Habitat Conservation Plan, 125–28
Heacox, Ed, 84–86, 98
Helicopters. *See* Seeding, aerial
Henderson, Howard, 26
High yield forestry program, 82–94
Higinbotham, Dean, *2*, 40, 41, 78–79
Hill, James J., 11
Homestead Act, 10
Hudson Bay Company, 7, 16
Hunt, Howard, *72*–73, 86, 92
Hunting, 67, *68*, 100–102
Hurricane Freida, 75–76

Ingram, Charles, 29, *30,* 31
Ivers Peak Fire, 79–81

Japan, export to, 76, 139
Jefferson, Thomas, 6, 9, 17–18

Kalahan, Clyde, *70*–74, 92
Karlen, Art, *45, 51*–52, 55, 69
Kelley, Hall J., 7–8
Kelly Lookout, *37,* 41–42, 45–46, 78–79

Land Ordinance, 18
Larsen, Jack, 127–28
Lee, Jason, 8
Leopold, Aldo, 95
Lieu Land Act, 12–13
Loggers, description of, 52–53
Logging, 48–53
Long, George, 13
Louisiana Purchase, 6–7

Manifest Destiny, 7
Mattson, Dan, 20, 48, 94, 96
McCullough, Rex, 134
McKeever, Don, 31
McLeod, Alexander, 16
McLoughlin, John, 7, 8, 9
Millicoma Tree Farm, dedication of, 44–46
Mills: CBX mill, 111–12, 123; closing of, 109–112
Morgan, Harry, Jr.: as assistant to Art Karlen, 68, *69;* as assistant to Bill Gaines, 55–56; on safety, 56; leaving Coos Bay, 72; presenting high yield forestry program, 86; under George Weyerhaeuser, 84
Mosquito Gulch, 62–63

National Environmental Policy Act (NEPA), 127–28
Native Americans, 3, 4, 17, 132–33. *See also* individual tribe names

Oregon: exploration of, 6; immigration to, 7–10, 17–26
Owl (Northern spotted owl), 105, 113–19, 122–23, 133–34

Pinchot, Gifford, 43, 44, 96, 97
Planting, 107. *See also* Seeding, aerial

Railroad (Northern Pacific): building of, 33–35; cruisers, 22; land grants, 10, 11–12; planning for, 31–32
Rainier National Park, 12
Roads, building of, 36–39, 61–62, 71
Roosevelt, Franklin D., 98
Roosevelt, Theodore, 13, *96,* 97, 134

Schenck, Carl Alvin, 45–46
Second American Forest Congress, 96–97, 134
Seeding, aerial, *87*–90, 92
Sierra Club, 133
Sommer, Herm, 40, 57, 68–69
Surveying, 17–20
Sustained yield, 82, 98

Tellico Dam, 115–16
Thomas, Jack Ward, 118
Timber and Stone Act, 10
Timber cruising, 22–23
Timber Culture Act, 10
Trapping, 101
Tree farms, 43–44, 98. *See also* Clemons Tree Farm, Millicoma Tree Farm

TVA v. *Hill,* 115–16

U.S. Fish and Wildlife Service, xv, 114, 115, 17, 127
U.S. Forest Service, 96

Weed, Oscar, 92–*93*
Weyerhaeuser, Dave, 31, 44
Weyerhaeuser, Davis, 96–97
Weyerhaeuser, Frederick, 11, 96, 97
Weyerhaeuser, George, 52, 84, *85*
Weyerhaeuser, J. P. "Phil," 32, 34, 50
Weyerhaeuser Company, xv: agreement with Fish and Wildlife Service, 127; description of, 54–56; and Forestry 21, 121–22, 124; land given to, 12–13; logged-Off-Land Department, 44; and Project Legacy, 124–25; purchasing lands, 26; recycling program of, 124–25; success of, 120. *See also* High yield forestry program
Wickendoll, Don, 93, 109, 110
Wight, Howard, Jr., 116–17
Willamette Meridian, 18, *19*
Whitman, Marcus, 8
Wolff, Jack, 93

Yale Forest Forum, 133–36
Young, Ewing, 8
Youst, Lionel, 26, 101, 102, 103

Library of Congress Cataloging-in-Publication Data
Smyth, Arthur V., 1919–
The Millicoma: biography of a Pacific northwestern forest / Arthur V. Smyth.
p. cm.
Includes bibliographical references (p.).
ISBN 0-89030-058-5 (pbk.)
1. Millicoma Tree Farm (Or.)—History. 2. Weyerhaeuser Company—History.
3. Forest management—Oregon—History. I. Title.
SD436.O7 S69 2000
333.75'097952—dc21 00-050372